DATE DUE

APR 29			

A serious but not ponderous book about

Nuclear Energy

Walter Scheider

Cavendish Press
ANN ARBOR

Cover photo: nuclear reactor plant
Cover graphic: from a 1950's brochure advertising uranium mine tours

Published by Cavendish Press Ann Arbor
Box 2588, Ann Arbor, MI 48106
e-mail at cavendish@worldnet.att.net
book orders and further information on this and other books, ideas, and materials for the learning and teaching of the sciences may be found at
www.cavendishscience.org

Publisher's Cataloging-in-Publication
(Provided by Quality Books, Inc.)

Scheider, Walter
A serious but not ponderous book about nuclear energy /Walter Scheider. -- 1st ed
p. cm.
Includes bibliographical references and index.
ISBN: 0-9676944-3-4 (hard cover)
ISBN: 0-9676944-2-6 (paperback)

1. Nuclear energy--Popular works. I. Title.
II. Title: Nuclear energy

QC792.4.S34 2001 539.7
QBI01-200080

Preface

Matter is not converted to energy. The equivalence of mass and energy is involved in fundamentally the same way in chemical fuels as in nuclear fuels. The equivalence of mass and energy is neither an explanation of how these fuels release energy, nor is that equivalence even necessary in order for the release to occur.[1]

When the complexities are stripped away, it's about bonds. Bonds are not the sole property of chemists. Bonds, whether chemical or nuclear, occur whenever a force of attraction brings two objects together and holds them there. In coming together, Work is done, and energy is released.

The nuclear force brings nuclear particles together, forming nuclear bonds, and energy is released. Along with that energy comes the mass of that energy.

Both the energy and the mass that comes with it are much greater in nuclear reactions than in chemical reactions. While the mass goes unnoticed in the chemical process, it is very much evident in the results of nuclear processes. So much so that it was in great measure the means for gaining information about nuclear reactions, which are not terribly well suited for laboratory experimentation.

On August 6, 1945, an atomic bomb small enough to fit in most living rooms, destroyed the Japanese city of Hiroshima. The reporters wanted to know how to explain this new weapon to the public. The technical people who briefed the reporters told them

---

[1] No where has this point been made as succinctly as in a note by Hanna Goldring, of the Weizmann Institute of Science, in *The Physics Teacher*, May 1984, p318

that because of Einstein, a way had been found to turn little tiny bits of matter into enormous amount of energy. So that's what the newspapers went with.

Nuclear science became immediately cloaked in the aura of relativity theory. The alchemy of turning tiny bits of matter into incredibly destructive energy was believed not because it was understood, but because there was a bomb, a city had been destroyed and a war concluded because of it.

At least one of the reasons that this confusing explanation was let loose, to become lodged in common wisdom and in text books, is that many otherwise competent scientists believed it. For so long had they been deriving energy predictions from mass measurements that they began themselves to think that the loss of mass caused the release of energy.

In the darkness, the cart had for all the world come to look as though it was in front of the horse.

Quite apart from failing to explain nuclear fission, the release of mass along with energy is not in the least unique to nuclear processes. It occurs wherever some source fulfils a need for energy in our world. Wherever and however we obtain a Joule of energy, there is (1 Joule/c^2) of mass that comes with it, and therefore goes missing from the fuel that provided us that Joule.

It is entirely ordinary and unremarkable that it should also occur in nuclear fission.

It is the aim in this book to straighten out this confusion, because only then can readers understand the way nuclear reactions provide energy. In the process we hope to help lift the veil of mystery and mystique that has turned so many among us into aliens in a field that is as fascinating as it is important to our lives.

This book also deals at length with practical questions of nuclear technology, the design of reactors, the nature of radioactivity, and the facts behind the assessment of risks and the arguments of policy. It also treats such matters of nuclear science as the chain reaction, the theoretical origin of the instability, and the fortuitous characteristic of the fission reaction that enables us to engineer it to a condition of stability.

****

Choices have to be made in preparing a book such as this that is designed for the inquisitive lay reader. These choices may leave some critics feeling that too much has been omitted or that some detail and some rigor have been slighted. It may at the same time leave other readers feeling that the goals are too ambitious, and that too much has been demanded of their capacity for abstract thinking and algebraic concepts.

With apologies to both these camps, we have decided to follow a path that has proved successful. This course of study has been a part of the author's physics course for twenty years, during which it has undergone modification and development in response to student reaction. It has required of the students nothing more in their mathematical repertoire than an introductory high school course in algebra

We have stayed with the Force-Work-Energy model of physical events that has been replaced in the quantum domain by the Interaction model. Yet even the most sophisticated treatments of the nuclear energy question invoke binding energies based on the forces of attraction between nucleons. Although the argument could be made that this skirts some issues involved with the events in a nucleus, a much stronger argument can be made for its pedagogic value and its essential correctness.

We have chosen to add to the usual descriptions of nuclear reactor design some quantitative insight into the questions of stability and control, and to dwell at some length on the remarkable properties of exponential functions which play such a central role in the harnessing of the chain reactions in nuclear energy production. We deal in plain language with the haunting question, "Why is the taming of the nuclear reaction so impossibly precarious?" and with the obvious next question, "How can we do it anyway?" These are vital bits of science and engineering for anyone looking to understand some of the issues of risk assessment attendant to choosing an effective public policy in the nuclear field.

****

It is only a short step from believing that "only the experts can understand nuclear physics," to accepting the conclusion that decisions and policies about nuclear technology can best be made by experts.

It is always bad government to leave any important public issue permanently and systematically in the hands of professionals, who, smart as they might be, have, by definition, a bias that may not be in the public interest.

Nuclear issues are life and death issues for all people on the globe. We are all at risk if decisions on such issues are based solely on the vested interest of those whose personal livelihood is involved with the very industry whose business it is to develop and build nuclear gear.

This is not to say that the views of experts and those in the industry are bound to be wrong, or that we should ignore the expertise and experience of professionals. It does say that it is vital to educate and empower a citizenry that can make its own informed judgments after listening critically to the experts.

To do otherwise is as foolish as it is, in the words of newspaper columnist Russell Baker, to leave economics to the economists.

Of course, as in most other disciplines, full and complete technological competence requires advanced study and advanced methodology. In this instance that would involve more elaborate use of mathematics and a more advanced familiarity with physics.

But a remarkably accurate and sensible understanding of the nuclear field is available to the average person willing to meet a book like this half-way. That means a commitment of reasonable time and effort, and a willingness to add and subtract and read simple algebraic language.

Beyond the practical, there is an intellectual treat in store for the reader. In gaining an understanding of the ordinariness of nuclear processes, one encounters once again that recurrent theme in basic science, the surprising commonality of fundamental ideas that span the chemical, biological, and physical world at the most elemental level. There is ultimately a single process by which fuels give up energy for our use, and this single process reaches from the burning of fossil fuels, to the generation of electric power from waterfalls, to the instantly ready energy stored in the muscle of frog, elephant, and human, and to the way in which energy is harvested from the atomic nucleus.

Nuclear Energy

Contents

PART I. The Source of Nuclear Energy

1. The truth about $E = mc^2$
2. Where do they shave off the little bits of matter?
3. Nuclear bonds – How much energy?
4. Energy from re-packaging nucleons

PART II. The Technology of Nuclear Energy

5. Activation
6. How to keep the log burning
7. Designing a bomb
8. Generations: Mothers and daughters
9. The squeeze between boom and bust
10. Nature's nuclear thermostat
11. Reactor architecture I: The Core
12. Reactor Architecture II: Driving the generator with nuclear heat

PART III. Other Reactors

13. Other coolant, other moderator
14. The Plutonium Reactor
15. Fusion

PART IV. Radioactivity

16. Radioactivity: What it is
17. How radioactive?
18. Radioactivity and Exposure

PART V. Risk assessment

19. Getting the information

APPENDIX. Minute by minute, hour by hour

Account of the Accident, from the President's Commission

Bibliography, Index

Nuclear Energy

PART I

The Source of Nuclear Energy

1.
The Truth about $E = mc^2$

The general public found out about nuclear energy in the evening newspapers August 6, 1945, the day that a nuclear bomb had been exploded over Hiroshima, a city in Japan, almost totally destroying it. Quoting scientists, the papers described what they called "atomic energy" as a brand new energy-yielding process, in which matter is not burned, but "converted" to energy, with unheard of efficiency.

The story was fed to reporters in the throes of the momentous events of the day. Military security had demanded that no advance word of the advent of the atomic bomb be in the hands of the media, so no advance preparation was possible to give adequate background to readers.

This is how the folklore about the source of this energy began. It has been passed down the generations since, causing more confusion than enlightenment. The story places the cart before the horse, confusing cause and effect. It takes an effect that is common and universal and makes it appear to be unique to the atomic nucleus, and in doing so obscures what actually *is* special about nuclear energy. It explains little, and leaves very important questions unanswered.

The gist of the folk tale is that the energy of the nuclear explosion comes from a "conversion" of the mass of small fragments of the nucleus into energy ("conversion" is in quotes because there is no conversion). It does not.

The story links this enormous nuclear energy release to the Einstein principle of the equivalence of mass and energy and the now-famous Einstein equation,[1]

$$E = mc^2 \qquad [1.01]$$

suggesting that, but for this equivalence principle, these little fragments of mass in the nucleus would remain forever locked there as matter, and the energy would never emerge. It has engendered hapless and frustrating bewilderment as to the nature of these little fragments of mass, where and what they are, and what, after all, impels them to undergo this Jekyll-Hyde transformation.

The historical role of $E=mc^2$

The concept of the equivalence of mass and energy, and the equation, $E=mc^2$, did indeed play a vital role in the work that led to nuclear energy release. It suggested a way to interpret some puzzling data that had been accumulating over many decades, that led ultimately to the discovery that indeed there is a large amount of energy involved in forming the bonds that hold the nuclear particles together. The capacity to release some of this bond energy lay in the nuclear particles like a tightly compressed spring ready, under the right conditions, to uncoil and leap out in a great fury, in a form that we can use, for good or evil, to nourish or destroy.

Nuclear bond energy is energy all along. It begins as energy and ends up as energy, and does not need to be converted to energy. Energy and mass are equivalent, in the ultimate sense that they are identical and are co-present always, two aspects of the same thing. And so this bond energy has as well, simultaneously, all the

---

[1] In this equation, c is the speed of light (3×10^8m/s), m is mass, and E is energy

properties of mass. Just as it begins as energy and ends up as energy, it begins also as mass and ends also as mass. This is true for chemical energy as well.

One of the properties of mass is that it has *weight* when it is in a gravitational field. And so, when the explorers of the periodic table weighed the atoms, they weighed the bond energy as part of the atom. When the force of attraction among the protons and neutrons creates new bonds in a new nucleus, bond energy is released, leaving less energy behind and the total weight of those nuclear particles diminished, so much that laboratory scales can detect it.

If you find yourself at this point content not to pursue questions about the source of nuclear energy, you may skip to part II dealing with nuclear technology – how bombs and reactors are built.

We don't recommend it, but if you choose to do so, you will find that you can learn about the technology without fully understanding the fundamental energy process in the nucleus. You can come back at a later time to deal further with the questions of Part I.

About bonds and bond energy

Before we go much further it is necessary to clear up one thing about bond energy. Unlike the coiling of a spring, in which the coiled spring has the energy, the bond loses its energy as the bond is formed. When the bond is there, the energy is gone.

This is not at all hard to understand. There are many kinds of bonds, and some are quite tangible and are part of every person's experience. What we say here about bonds is not specific to any one kind of bond, but is totally general about all bonds.

A *bond* is the name we give to what pulls two things toward each other and holds them there. It requires a force of mutual attraction to form a bond.

Consider the gravitational bond between you and the earth. Although it may seem that the earth does all the pulling, in fact you and the earth attract each other (and with equal force). No matter how high you climb on this earth, the bond between you and the earth is there. But when you jump down from a height, that bond between you and the earth tightens, releasing energy.

How does it do that? Here is how: The force of attraction causes you to go faster and faster, accumulating, as you fall, an amount of kinetic energy. That portion of energy is no longer part of the potential energy of the bond; it is *yours* to do with as you please. When you land, you have to do *something* with it, because you must stop. The collision can turn it into heat. Or you can jump on a spring and compress it, let the spring store the energy. Or you can slide down an incline and give that energy up to friction as you slide. Regardless of where the energy goes, it is not there as bond energy when you are at the bottom.

The kinetic energy that water obtains by falling from a lake at a higher level down to the bottom of a waterfall can be used to generate electricity. This is the basis of hydro-electric power. When a bond is made or tightened, energy is released. The bond energy is there before the bond is made, and is not there afterward because it has been released.

Another example comes from the chemistry that runs your car. Gasoline is a fuel that can supply you with energy because there is a bond that it *has not yet formed* with the oxygen of the air. Do not forget that although you pay for only the gasoline, without the air that comes free, that gasoline could not give you energy. You are willing to pay for gasoline because the carbon and hydrogen in the gasoline molecule can bond with oxygen but *have not yet done so*. Put the gasoline and the air together, make a spark, and the gasoline forms bonds with oxygen, producing carbon dioxide and hydrogen oxide (water), giving you heat energy and exhaust gases.

The auto exhaust contains all the atoms of the gasoline and the oxygen, but now that there are bonds between the atoms, it is worthless as a fuel.

There are many other kinds of bonds. A stretched spring can pull two objects together, but in doing so, the spring gives up its "spring energy."

Resting muscle stores Adenosine tri-phosphate (ATP); when that muscle needs quick energy, one of the phosphates goes into solution, allowing the remaining Adenosine di-phosphate (ADP) to become the more tightly bonded molecule that it always wanted to be, in the process passing energy to your muscle fiber.

In the *formation or tightening* of *all* these bonds, *energy is released.* An experiment can be designed, usually without great difficulty, to determine *how much* energy is released. To find out how much is released with the burning of gasoline, take an ounce of gasoline, put it in a little camp stove, and burn it under a pot of water. You will find that you can boil about a pint of water.

Bond energy and mass

It takes about 280,000 calories to boil that pint of water. That energy came from the ounce of gasoline and from the oxygen that combined with it. More precisely, it came from the formation of the bonds during the burning of the gasoline.

280,000 calories of energy that were present in the gasoline and oxygen while they were separate, are no longer there in the carbon dioxide and water. Equation [1.01] allows us to calculate how much mass is the equivalent of 280,000 calories.[2] The mass of 280,000 calories is 0.000000013 grams.

The large value of c^2 ($9 \times 10^{16}\, m^2/s^2$) accounts for what seems like an absurd imbalance in the amount of mass and the amount of energy. Nonetheless, no process that yields energy is an exception: wherever energy goes, its mass goes also.

If instruments were available to measure the decrease of 0.000000013 gram that occurs when you burn an ounce of gasoline with oxygen, that measurement could be used to determine that 280,000 calories were released. This kind of reasoning is precisely what gave us the first clue to the amount of energy available in the formation of nuclear bonds.

Nuclear force, nuclear bonds

Early in the 20th century, when it was found that the atomic nucleus contained positively charged unit building blocks, which came to be called protons, each a replica of the hydrogen nucleus, it became evident that there was a cohesion problem. With a lot of positively charged particles all in a very tiny space, the electrical forces of repulsion at such short distances[3] would tend to blast such a nucleus apart and scatter all those positive particles far and wide. There were no negatively charged particles in the nucleus to counteract the repulsive forces, and it was clear that a force up until then unknown, but obviously very strong, at least at short range,

---

[2] If you do this calculation, you must remember to convert from calories to Joules (1 cal = 4.18 J) so that the result will be in kilograms.

[3] Coulomb's Law has the square of the distance of separation in the denominator

had to be present. This force is still now called the "Strong Nuclear Force."

This force is called a "nuclear force," not because there is anything particularly nuclear about it, but because it is a very powerful force whose range of action is very short, so short that only at distances within an atomic nucleus is anything near enough to be subject to this force. The nucleus is extremely small, even compared to the size of the smallest atom. If a model were built of an atom, in which the atom as a whole (with its electrons) is made the size of a large football stadium, the nucleus would be approximated by a pinhead on the 50 yard line. In numbers, atomic diameters are of the order of 10^{-10} m, while nuclear diameters are about 100,000 times smaller, of the order of 10^{-15} m.

But the strong nuclear force, or "Strong Force" for short, is extremely strong over its range of effectiveness, and in the formation of nuclear bonds can do an enormous amount of Work, even though the movement of the attracting particles is very short.

The Strong Force, obviously is providing forces of attraction of sufficient magnitude within the nucleus, to hold it together. In the 1930's the neutron was discovered, and ultimately it was found that the Strong Force does not know the distinction between protons and neutrons.

The word, "nucleon," was coined that applies to both protons and neutrons, just as, the word "people" applies to both men and women. The Strong Force, then, is a force between any two nucleons.

The implication is that if you take two protons and two neutrons and put them together to make a helium nucleus, the Strong Force would weld some good strong bonds among those four nucleons to hold them together against the electric repulsion between the two protons. In the formation of those bonds, a certain amount of energy is released.[4]

How much energy?. No simple experiment like boiling water over a small one ounce gasoline flame can be done to measure the bond energy released in the formation of these nuclear bonds.

The Mass Defect

It is here that an old puzzle and a new theory came together to offer an answer.

The old puzzle was that there appeared to be a flaw in the early view of the meaning of the periodic table. As the data came in to fill the blanks in the periodic table, it had at first appeared that the atomic masses came out as integer multiples of the mass of the smallest, hydrogen. If hydrogen is assigned the atomic mass of 1, Helium had an atomic mass of 4. Lithium had an atomic mass of 7. Beryllium, 9. Boron, 11. Carbon, 12. Nitrogen, 14. Oxygen 16, and so on.[5] But as more precise data arrived, it turned out that the ratios were only approximately integers. The ratios, however, were so close to being integers that the idea that atoms were built out of identical building blocks, each of mass equal to that of a hydrogen atom, was not abandoned.

The small deviations from integer multiples were most troubling. They did not appear to be random and exceptional, but in fact had

[4] The energy released is a net amount, that is left over after the repulsion force between the protons is canceled out.

[5] A few deviated substantially from the integer multiples pattern, giving rise to an early concept of isotopes, elements that contained different sub-species, each with a different integer number of building blocks. The atomic mass of an element might then be an average of two isotopes. Chlorine, for example, with an atomic mass of 35.5 has turned out to be three fourths Chlorine-A with 35 building blocks, and one fourth Chlorine-B with 37 building blocks.

a somewhat regular pattern, increasing steadily with heavier elements.

For example, the atomic mass of helium came out to be just about 3.97 times that of hydrogen. The atomic mass of oxygen was found to be 15.87 times that of hydrogen. There appeared always to be something missing. Is it possible that just the mass of the building block was slightly off? Apparently not; the deviations did not disappear if the mass value of hydrogen were adjusted. Quite apart from hydrogen, if you took three heliums and made a carbon out of it, the mass of a carbon atom was not three times, but only 2.998 times, the mass of a helium atom. The seemingly missing mass was called the "Mass Defect."

We now know that this mass defect is due to the release of bond energy, something that occurs throughout all energy-yielding processes. When the mass defect was observed in the periodic table, it seemed a unique phenomenon. No similar defect had been observed anywhere else. Combining an ounce of gasoline with oxygen gives a mass defect of 0.000000013 grams.

The mass defect had not been observed in chemical reactions, not because it isn't there, but because it is too small to measure. But if we take an ounce of hydrogen, and "glue" the atoms together four at a time, making helium atoms out of that ounce, we get not 1.00000 ounce but 0.9929 ounces of helium. The missing mass weighs 0.0071 ounces (the weight of a mass of 0.200 grams). This amount can be determined by weighing, which is exactly how it was determined. The atomic masses were determined using chemical balances.

No one can go into the laboratory and make helium out of hydrogen, although the sun can do it. The mass defect was determined from separate measurements on hydrogen atoms and a comparable number of helium atoms, which could be done.

Not for several years after Einstein first published his theory of special relativity did he discover that one of its implications is the equivalence of mass and energy. With that insight, the mass defects in the periodic table took on a possible new meaning, that they represented an astoundingly large "energy defect." The helium mass defect of 0.200 grams is the mass equivalent (using [1.01]) of 1.80×10^{13} Joules, or 4.3×10^{12} (that's 4.3 trillion) calories. The large value of c^2 that made the mass difference come out so small in the example of gasoline, now makes the energy yield very large in the example of helium.

Could this be? At first few gave much credence to that possibility. In fact the whole idea of the equivalence of mass and energy struck even Einstein as puzzling. When he discovered that his postulates of relativity seemed to imply this equivalence, he confided to a friend at the time that, "...this thought is both amusing and attractive, but whether the Lord laughs at me concerning this notion and has led me around by the nose – that I cannot know."[6] When this equivalence first came to his attention, there was no experimental evidence to confirm it. Now, almost a century later, there is an abundance of experimental confirmation of it. There is no longer any need to be concerned that he may have been "led around by the nose."

This clue to the nature of the mass defect in the periodic table led to intense activity, making calculations, and considering the implications.

The first naive idea of the equivalence of mass and energy is that it is a "conversion recipe," that shows how to "make energy out of matter," just as you make a cake out of sugar and flour. But this naive view maintains that at any one time it is either one or the other, matter or energy, cake or ingredients. Those who clung to this image long after Einstein abandoned it in favor of a "co-existence principle," had a difficult question to deal with. Many of

---

[6] letter to Conrad Habicht, widely quoted, eg, in Hey and Walters, "Einstein's Mirror," Cambridge University Press, Cambridge, UK, 1997.

them, including excellent writers on nuclear physics,[7] never quite came to grips with the problem of where those little bits of matter, those 0.200 grams of nucleus stuff, are in the atom and what they look like before they are "converted."

Problems

1.1 Show that the 280,000 calories of energy released in burning one ounce of gasoline have a mass of 0.000000013 grams. Use unit conversions: 1 kg = 1000 grams; 1 calorie = 4.18 Joules; 1 Joule = 1 $kg\,m^2/sec^2$.

1.2 Recall that when it *requires* energy to be put in to break or stretch bonds, and energy is *released* when bonds are made or shortened. In the following pairs, tell which weighs more (because it has more energy and therefore more mass):

A stretched rubber band	An unstretched rubber band
hot coffee	cold coffee
gasoline and oxygen	exhaust gases from same
an ice cube at 0°	water from the melted cube at 0°
iron ore (iron oxide)	the same iron and oxygen after the oxygen has been driven out in a blast furnace.

1.3 Find the force between two protons (charge = 1.6×10^{-19}Coul) separated by a distance of 1×10^{-15} meters (the size of an atomic nucleus). Use Coulomb's Law:

$$F = (9\times10^9 \text{ Newton m}^2/\text{Coul}^2)\times Q_1\times Q_2/r^2$$

Q is electric charge; r is the distance between the charges.
Use Newton's Law, F = ma, to find each proton's acceleration.
Mass of proton is 1.67×10^{-27} kg.
What does this tell you about the magnitude of nuclear forces?

---

[7] including the authors of some of the best books on nuclear energy, such as Paul Craig and John Jungerman, who worked on the atomic bomb project (see Bibliography).

2.
Where do they shave off the little bits of matter?

It's a fact: the protons and neutrons get smaller. Not necessarily smaller in size, but smaller in mass. Even the molecules of gasoline and oxygen get smaller.

It is very hard not to ask, where did they shave it off? Our concept of electrons and protons and neutrons is that they are what they are, and there aren't some that are bigger and some that have been shaved.

In this chapter we will try to come to terms with what "co-existence" of mass and energy means, and then answer the question, where did they shave off the stuff that is both mass and energy?

"Co-existence" of mass and energy does not mean that there is some of each, side by side. It means that whatever has the properties of energy also has the properties of mass, always has, always will, and doesn't turn from being one to being the other. It helps to think of two sets of "properties," because we are familiar with objects that have multiple properties. A ball is red and round. The weather is cold and sunny.

Imagine that a person with hearing who is not sighted, a person who is sighted but can not hear, and one with both sight and hearing, attend a performance of Tchaikovsky's ballet, *The Nutcracker Suite.* One reports the spirited rendition of the melodies, the reverberant fullness of the orchestra, the precision of the strings. Another tells of the fabulous costumes, the energy and perfection of the dancers, the splendor of the scenery, the stunning lighting effects. Would not the person with sight and hearing maintain that both sets of characteristics are there, not one at a time, but both at the same time? Why would anyone insist that there had been a conversion of the one set of characteristics into the other?

Mass and energy look very different to us. It is no wonder that even Einstein was surprised to find that his special theory of relatively implies their equivalence.

If we are trying to come to terms with the fact that mass and energy are not one thing at one time and another at another time, but are both simultaneously, always and forever, then we must do so in terms of their properties.

It is understandably hard to imagine that something is two things at once if one has only the vaguest idea of what either one "is." What something "is" is defined, in physics, by its "properties," or "characteristics." What a physicist means by saying "*this* is an electron," is that *this* has the *properties* of an electron. It is considered uncool to talk about something, if that something does not have a specific, accepted, definition in terms of properties.

The properties of mass; the properties of energy

Mass has two properties: (1) Inertia, the sluggishness that is described by Newton's Law of Motion, that makes an object of large mass more difficult to accelerate. And, (2) Weight in the presence of a gravitational field.

Energy is defined in a somewhat more complicated way. Because "energy" is a household word, we have an intuitive sense of what its properties are. Physicists require a more specific description, even at the cost of being a bit more obscure. The physicists' definition includes some common usage, as in, "We used 300 KWhr of electric energy this month, and are paying the electric company for that energy." It does not include some other common usage, such as, "I'm putting way too much energy into pleasing my boss."

Energy is defined in classical physics by the *Work–Energy* Theorem, which includes a definition of *Work* in terms of the movement of an object during the exertion of a force on it. The *Work–Energy* Theorem is usually written as an equation that contains a number of terms. Kinetic Energy (the energy of motion) is one term, and then for each of the forces that are known, there is a term for the *Work* done by it. The statement embodies a conservation law that says the total of the changes in these terms must be equal to zero during any process. The Greek letter, Δ, means "Change in..."

$$\Delta(\text{Kinetic Energy}) - \text{Work done by all forces} = 0 \qquad [2.01]$$

Work done is always done at the expense of potential energy.[1] For each Work term there is a change of potential energy. Bond energy is potential energy associated with the force that makes the bond.

$$\Delta(\text{Kinetic Energy}) + \text{total } \Delta(\text{potential energy}) = 0 \qquad [2.02]$$

If there is to be a gain in kinetic energy (which includes both large scale motion and the random motion of heat), there must be a corresponding total amount of *Work* done, meaning, potential energy expended. It means, quite generally and universally, that to get energy from a fuel, *Work* must be done, and that means that a force has to move something. The forces of nature are either forces of attraction or forces of repulsion. When a force of attraction

---

[1] Work done by most forces is expressed as a change in potential energy. Terms for Work done by so-called "non-conservative" forces are temporary substitutes for the sometimes difficult assignment of this Work to potential energy terms. "Muscle force" is an example that may appear as a "non-conservative force." Ultimately, when muscle force is broken down into the chemical changes that release this energy, this term can be replaced by potential energy changes. This may be very complicated to do computationally, and may for that reason be avoided. But this does not alter the generality of the statement made about the terms of the Work Energy theorem.

moves objects toward each other, a *bond* is formed or strengthened,[2] at the expense of *bond potential energy*.

Bond energy has the properties of energy, meaning that its expenditure can cause an increase in heat energy. It also has the properties of mass, meaning that it has weight, and adds to the sluggishness of the particles involved.

This anatomy of bonds that we are so curious about is easier to see in examples in which the bonds are electric force bonds, so let us look at three examples of electric force bond energy.

The anatomy of bonds

Example 1: Electric Force Bond between ions

It is clear from earlier discussion that the bond energy is not in the bond. The bond energy is there before the bond is formed, or tightened. As the bond is made, the energy is released.

It is easiest to show this for the case of an electric bond (all chemical bonds are electric force bonds). We can generalize to nuclear bonds from that.

The bond of this example is one that is familiar to most chemistry students; they have made these bonds in their laboratories. It is the bond made during the neutralization of base with acid, for example the neutralization of hydrochloric acid with sodium hydroxide solution.

A bond is formed between a positive hydrogen ion (H^{+1}) and a negative hydroxyl ion (OH^{-1}). Each ion is initially surrounded by an electric field. Each isolated ion has an electric field that stretches

---

[2] In the rarer cases where *Work* is done by a force of repulsion, as in the ejection of an alpha particle from a radioactive nucleus, two things that are held together against their "will" are released to fly apart

far. The electric field, E, at a distance "r" from the center of a charge, "q", is given by the Coulomb equation,

$$E = (1/4\pi\varepsilon)\times q / r^2 \quad [2.01]$$

where ε is the dielectric constant of the surrounding space. The electric field points outward from positive charge and inward toward negative charge.

In Fig. 2.1 are the two separate charges, the negative hydroxyl ion on the left, consisting of an oxygen with one hydrogen already tacked on, and a hydrogen ion on the right. Each is surrounded by its own Electric Field.

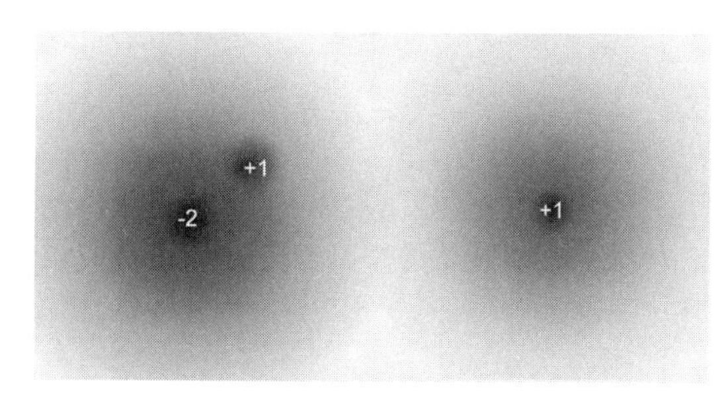

Fig 2.1 A hydroxyl and a hydrogen ion, with their electric fields.

The two ions are of opposite charge, and attract each other. The force of attraction causes both to move in the direction of the force, bringing the ions together. They move towards each other until the hydrogen ion breaks loose of its links to the water it is dissolved in and forms a new bond with the OH^- ion.

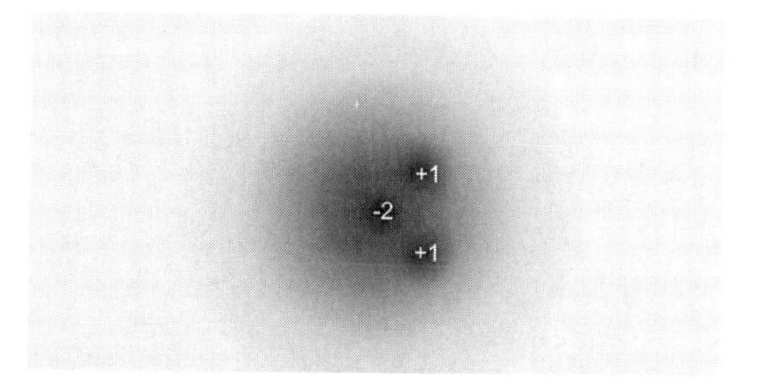

Fig 2.2 The electric force of attraction has formed a bond between the ions. The electric field is shrunken, and decreases sharply with distance because the total charge is now zero.

Neutralization has produced a water molecule (Fig 2.2). The effect of the bonding is exaggerated, and the illustrations are not drawn to scale.

Fig 2.2, diagrams the diminished electric field at a distance. Seen from far away, the water molecule (H–OH, sometimes written H_2O) has zero total charge, and looks like an uncharged particle.

The neutralization of one liter of 1-Normal hydrochloric acid with one liter of 1-Normal sodium hydroxide yields 13300 calories, enough to raise the temperature of the mix by about 6 degrees.

Bond energy was released in the formation of the bonds. As long as the mixture of the solutions is still warm, the energy is in the mixture as heat. As the mixture cools back to room temperature, the heat energy produced by neutralization leaves, and takes with it, based on $E=mc^2$, about 3.5×10^{-14} grams of mass. To measure this change would require 17–decimal place weighing accuracy. Not likely. But, if it *could* be measured, the mass decrease would be found to be 3.5×10^{-14} grams.

Where did the *mass* of this bond energy come from? 3.5×10^{-14} grams doesn't seem like much, but, on the other hand, it is the mass of about 1.17 billion water molecules. Did 1.17 billion water molecules get "turned into" energy? The answer is, "No," because one can neutralize just one hydrogen ion and one hydroxyl ion making one water molecule. Each pair of ions that are joined in a bond has to contribute its share of the mass that is released.

Where had the mass been that disappeared during the formation of the H–OH bond? All the oxygen and all the hydrogen are still there. Only the electric fields appear to have been affected.

It has long been known that there is energy associated with an electric field. For each little microscopic cube of space in which there is an electric field, one can calculate an amount of energy that the field in that little cube contributes. The energy in such a little cube divided by the volume of that cube is called the "Energy density" of the electric field.

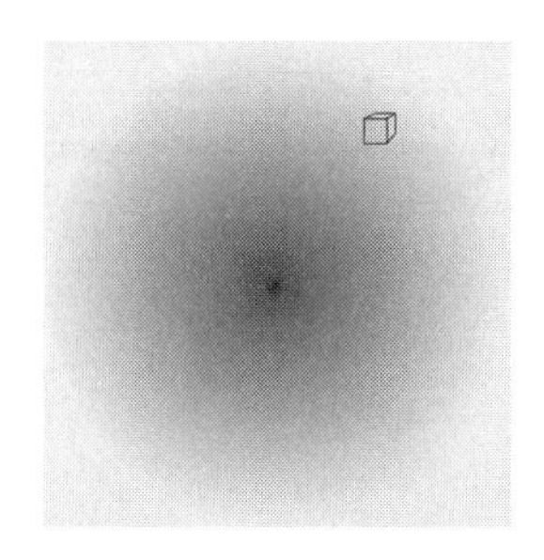

Fig 2.3 In every cube of space in an electric field there is some electric field energy. This energy is also mass.

The energy density is proportional to the square of the magnitude of the electric field. The equation for the energy density, "*u*," (measured in Joules per cubic meter) is,

$$u = \tfrac{1}{2}\varepsilon E^2 \qquad [2.03]$$

(ε is the dielectric constant. If the electric field is in empty space, the universal constant, ε_0, is used for ε)

As the ions went from being separate (Fig 2.1) to being together in the water molecule (Fig 2.2), the total energy in the electric field decreased. The energy was released at the expense of electric field energy.

And so it seems that the electric force bond energy is electric field energy!

But energy is also mass. And so, the energy that has been associated with the electric field, has a mass equivalent. This means that the mass released was electric field energy.

The anatomy of bonds

Example 2: Bond between proton and electron

Instead of two ions, let us now deal with two elementary particles, in empty space: a proton and an electron. Each, separately, has an electric field surrounding it. The fields of the separate proton and electron (Figs 2.4 and 2.5) look similar to the fields of the ions of the last example; in both cases the fields are spherically symmetrical, meaning that they look the same from any direction. The fields of the proton and electron are represented in these diagrams by "field lines" that radiate outward from the proton and inward toward the electron. The density (how close the lines are) roughly represents the value of E, the field intensity. As distance from the particles' centers increases, the lines spread farther apart, indicating that the field intensity gets smaller.

Fig 2.4 A proton and its electric field.

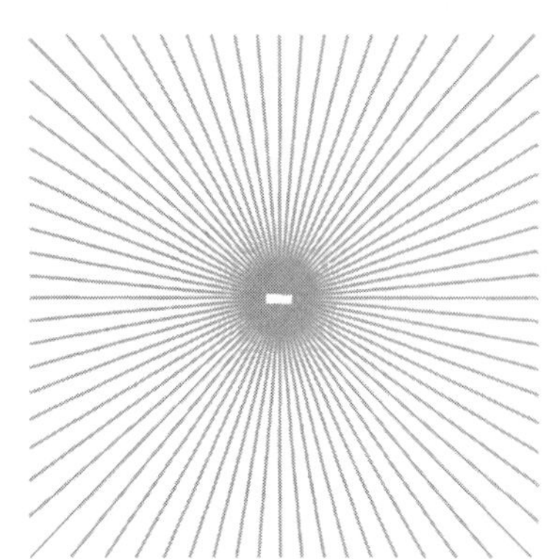

Fig 2.5 An electron and its electric field.

Letting a loose electron and a loose proton join up in a laboratory experiment is something that we know how to. The end result of the experiment is a proton surrounded by the electron in a spherically symmetric pattern: a hydrogen atom.

The apparatus is a hydrogen gas discharge tube, familiar to most chemistry and physics students. When the tube is connected to a high voltage source, hydrogen atoms are torn apart into their separate protons and electrons. Then, because they really like being together better (it's a lower energy state) the electron "falls" back into its place with the proton, re-creating the hydrogen atom.[3]

---

[3] What we describe here applies to atomic hydrogen. The diatomic hydrogen molecule, H_2, is somewhat different. In a gas discharge tube there is sufficient atomic hydrogen to give the spectral line described here.

The energy released as this bond is formed is emitted as a "photon," a little wavelet of light, which is electromagnetic energy. In the case of these electrons that "fall" to the ground state in a hydrogen atom, the light is far in the ultra-violet portion of the spectrum. (Of the many lines in the hydrogen spectrum, the human eye observes only the photons produced by electrons that land, temporarily, on a "ledge," waiting to drop the rest of the way in a subsequent jump.) Ultra-violet light is not visible to the eye, but can be "seen" with the aid of instruments. The line in the hydrogen spectrum that is produced by free electrons falling to the ground state directly is the line deepest into the ultra-violet, the line of shortest wave length of the entire hydrogen spectrum. Its wave length, λ, is 91.15 nm (1 nm = 1 nanometer = 1×10^{-9} m).

The theory of photons gives us an equation that relates the energy, E_λ, of a photon to its wave length (its color).

$$E_\lambda = hc/\lambda \qquad [2.04]$$

where "h" is Planck's constant, 6.62×10^{-34} Joule sec, and c is, as usual, the speed of light, 3×10^{8} m/s.

Putting 91.15 nm for the wave length, λ, tells us that the energy of the photon is 21.76×10^{-19} J.

From equation [1.01] we find that the *mass* of this photon is 2.418×10^{-35} kg. A rather tiny little thing.

We start with a proton and an electron, each separate, as were the two ions in the previous example. Since each has the same electric charge, except that the electron is negative and the proton is positive, their electric fields look the same, except that one points outward and the other points inward (Figs 2.4 and 2.5).

When the electron and the proton come together to form a hydrogen atom, what does the atom's electric field look like?

The electron-proton pair do not look like the ion pair in the first example. They do not sit side-by-side after forming the bond. The electron (which is sometimes pictured as "orbiting" around the proton) surrounds the proton in a puffy cloud described by quantum physics as a "probability density" function that permits calculation of the probability of finding the electron at various distances from the proton.

An electron is an elementary particle, that can not be sub-divided. "Probability of finding the electron" means just that: if you check at some instant, you may find *it* here, or you may find *it* there. But you don't find *some of it* here and *some of it* there. That makes it tricky to translate this probability function into a density function, a distribution in spherical layers, like peanut butter and jelly layers, although that is done, with limited legitimacy. The result can be thought of as a time average density in that electron cloud around the proton.

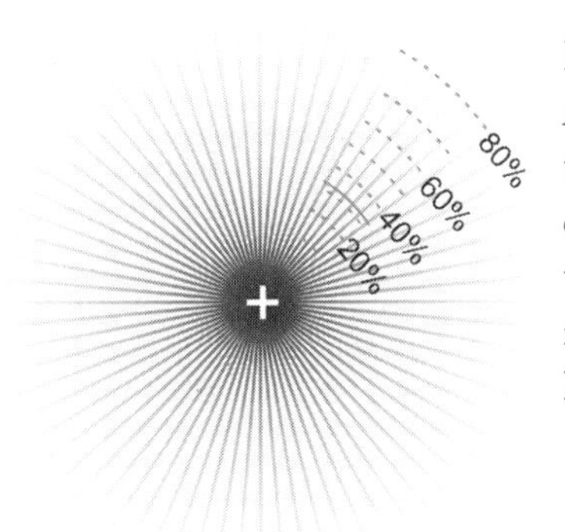

Fig 2.6 A hydrogen atom. Arcs mark at 10% intervals the total fraction of the electron's probability density that is inside a sphere of that radius. The solid arc is the location of the Bohr radius.

The "time averaged electron probability density" is greatest near the proton, decreasing and tailing off to about 13% of that density at a distance called the "Bohr radius," which is 0.529×10^{-10}m.

The arcs drawn in Fig 2.6 show 10% segments of the distributed electron's mass. The solid line marks the Bohr radius. The electron is obviously widely distributed, and occupies, over time, the vast majority of the space of the atom. The proton itself is about 1/100,000th of the diameter of the hydrogen atom, and could not be seen if it were drawn as a dot in the picture of Fig 2.6.

The lines representing electric field outward from the proton in Fig 2.6 fade out to indicate the effect of canceling electric field due to the presence of the negative electron. Because the proton-electron complex is neutral as a whole, the electric field diminishes to zero more rapidly than it did around each particle separately.

This makes an interesting portrait of strongly diminished electric field associated now with the hydrogen atom. All that can be said in the present context about the origin of the energy of the photon is that its mass equivalent is now missing from the proton and electron, and that the original electric fields of both can be re-established by supplying an amount of energy equivalent to that of the departed photon. This will re-establish the field and bring back not only the 21.76×10^{-19} J of energy taken away by the photon, but also the 2.418×10^{-35} kg of mass that went with it.

The anatomy of bonds

Example 3: Bond between two cookie sheets

The field around the hydrogen and hydroxyl ions is too complicated to do the numbers. The time averaged field that quantum physics tells us surrounds the proton in the hydrogen atom is not usable for energy calculation. In both cases we relied on qualitative arguments to conclude that the energy released came at the expense of the diminished electric fields.

It is time to take up an example that can be done exactly. It is a large scale example, easy to visualize, providing a convincing confirmation that the energy emitted in the formation of an electric field bond comes at the expense of the electric field. Recalling that chemical bonds are electric field bonds, conclusions drawn from these examples extend to the entire realm of chemical bonding. Nuclear force bonds will be brought under the same umbrella, because they are formed in a similar manner, although through the action of a different force.

The present example requires two flat, metal "cookie sheets" [4] that can be placed parallel and close to each other (in order to minimize fringe effects around the edges) and a battery to charge up the cookie sheets.

The battery will pump a certain amount of charge from one cookie sheet to the other, leaving the two sheets oppositely charged, one positive and one negative. The batteries must be disconnected once the sheets are charged, so that the charge, once put there, will remain there.

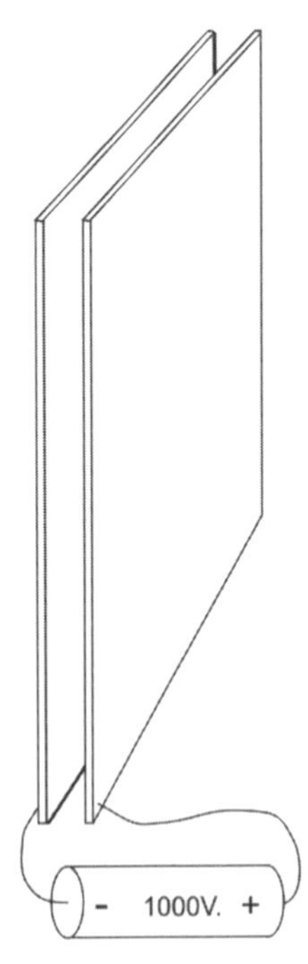

Fig 2.7 The cookie sheets

The cookie sheets, oppositely charged, will attract one another, as do all positive and negative charged objects. The desirable feature of this arrangement is that the electric field, E, in the space between the cookie sheets is uniform (the same everywhere in the space), and remains constant even while the cookie sheets move toward each other.

The electric force of attraction between the cookie sheets remains constant as they approach each other, as does the *energy density* of the field, but as the space between them shortens, the *total energy* of the field decreases.

[4] Any pair of flat metal surfaces will do.

We propose to use two cookie sheets each 1.0 meter square. We will place them parallel and 8.85 mm apart, just a little under a centimeter, or about 3/8 inch. We'll place them vertically, so that gravity will play no role in their movement toward each other. We'll attach one cookie sheet to an insulating frame, and hang the other by two thin threads from a high ceiling, so that it can move freely in the horizontal direction, but can't twist.

We will use a 1000Volt battery, or a power supply that can give us this voltage, momentarily. It would be wise to be careful around these charged cookie sheets, should you decide to actually do this experiment.

---

1. Calculation of the energy of the electric field that is lost

The charge, Q, that will appear on each cookie sheet is then 1.0 μCoul, (1 μCoul, "micro-Coulomb," is one millionth of a Coulomb), positive on one sheet, and negative on the other.

The electric field in the space is given by,

$$E=\sigma/\varepsilon_0,$$

where σ is the surface charge density, Q/(Area), on the cookie sheets, and ε_0 is 8.85×10^{-12} Farads/meter.) The area of each cookie sheet is 1.0 m^2, giving 113,000 N/Coul for the electric field.

The energy density of this electric field, is

$$u = \tfrac{1}{2}\varepsilon_0 E^2 \qquad [2.03]$$

The energy density, u, comes out to be 0.05650 Joules/m^3.

The total energy of the electric field between the sheets at the beginning of the experiment is the energy density times the volume, which is the space thickness times the area of the cookie sheets, or,

$$\text{vol} = (0.00885\,\text{m})\times(1\,\text{m}^2) = 0.00885\,\text{m}^3$$

Then,

Energy = $u \times \text{vol} = (0.05650\,\text{J/m}^3) \times (0.00885\,\text{m}^3)$

Energy = 0.0005 Joules, exactly

===

As the cookie sheets attract and move towards each other, the electric field remains the same, but the volume decreases until it is zero at the point where the cookie sheets crash into each other. 0.0005 Joules of electric field energy has vanished. Where did it go?

===

2. Calculation of the kinetic energy gained by the moving cookie sheet

Let us imagine that we hold one cookie sheet fixed, and let the other approach it. During the approach of the cookie sheets, the energy went into the kinetic energy of the one movable cookie sheet. The force on the moving cookie sheet is due to that part of the electric field that is produced by the other cookie sheet,

$$E_1 = \tfrac{1}{2}\sigma/\varepsilon_0 = \tfrac{1}{2} \times 113{,}000\,\text{N/Coul} = 56{,}500\,\text{N/Coul}$$

$$F = E_1 \times Q = (56{,}500\,\text{N/Coul}) \times (1 \times 10^{-6}\,\text{Coul}) = 0.0565 \text{ Newtons}$$

This is a force of about 0.20 ounces.

If the mass of the moving cookie sheet is 0.50 kg, it will experience an acceleration (from Newton's Law, $F = ma$)

$$a = F/m = 0.0565 \text{ Newtons}/(0.50\,\text{kg}) = 0.113\,\text{m/s}^2$$

The rules of kinematics tells us that the moving cookie sheet will be traveling towards the fixed cookie sheet with a speed of 0.0447 m/s (about 0.101 miles/hour) when they crash. Its kinetic energy is

Kinetic Energy = $\tfrac{1}{2}mv^2 = \tfrac{1}{2}\,(0.50\,\text{kg}) \times (0.0447\,\text{m/s})^2$

Kinetic Energy = 0.0005 Joules

===

The instant before the cookie sheets crash into each other, the entire 0.0005 Joule of electric field energy is gone, and the moving cookie sheet has kinetic energy of exactly 0.0005 Joules.[5]

What happens after they crash is beside the point. On crashing, that energy becomes mostly heat energy, which then can warm the environment. 0.0005 Joules of energy has been released.

The energy was given up by the electric field in the process of shortening the electric force bond between the two cookie sheets. Joule for Joule, exactly, it went first into the motion energy of the colliding objects, eventually into whatever we chose to do with it, such as heat the environment.

It is no mystery whatever where this bond energy came from. It came from the force field that produced the bond. It happened quite without any intervention on the part of $E = mc^2$. If mass and energy were *not* equivalent, this bond energy would still have been released, by exactly the same mechanism and for the same reason.

And now, what about $E = mc^2$? Well, of course, the 0.0005 Joules of energy released has a mass equivalent, 5.556×10^{-21} kg, not even measurable with available weighing instruments. Yet that 0.0005 Joules of energy did not leave without its 5.556×10^{-21} kg of mass. Nor did that mass come from little shavings off the aluminum sheet (even if invisibly small). Not one atom of aluminum is missing, and it would take 123,000 aluminum atoms to make up even that small mass. The mass came from the mass of the 0.0005 Joules of electric field energy, which, in the formation of the bond, had disappeared.

---

[5] When the cookie sheets touch, they discharge their electric charge, one into the other, but no energy is involved. The voltage between the cookie sheets has decreased to zero.

Nuclear bonds and nuclear energy

We have looked at three examples, the first two in appearance more like what happens on the nuclear scale, but the last one most dramatically and exactly tracing the energy released to the energy of the electric field. What happened in all three examples also happens in the nucleus, except not by action of an electric field, but by the action of a field related to the nuclear strong force.

The electric forces we have examined are weak compared to the nuclear strong force, and so the examples have dealt with mass equivalents of unmeasurably small magnitudes. The nuclear strong force is, at short range, a much stronger force. In producing or strengthening nuclear bonds, much larger amounts of energy are released – amounts whose mass equivalent *is measurable* with ordinary measuring devices.

Mass–energy equivalence plays a role in nuclear physics exactly because the mass equivalents of these bond energies *are measurable.* The measurement of the *weight* of the missing energy became the most convenient way to measure the *energy potential* of the nuclear reactions. Using the relation, $E=mc^2$, became a vital tool for obtaining numbers that otherwise were inaccessible.

You can find how much energy is released in burning an ounce of gasoline by lighting an ounce of gasoline, and measuring the rise in temperature of a small beaker of water set above the flame. This can't be done as easily with nuclear reactions. Yet the same question is there to be answered: How much energy is released?

Problems

2.1 Very close to a proton, the electric field is very strong.

(a) Find the magnitude of the electric field at a distance of 10 nuclear diameters, or about 1×10^{-14} m. Use Eq 2.01, with the value of ε being its vacuum value, $\varepsilon_0 = 8.85 \times 10^{-12}\,\text{Coul}^2/\text{Joule m}$. For q use the charge of a proton, 1.6×10^{-19} Coul. Your answer should be in units of Newtons/Coul.

(b) Find the energy density (Joules/m^3) using Eq 2.03.

(c) Find the surface area of a sphere of radius 1×10^{-14} m. Then imagine a thin shell of thickness 1×10^{-15}m at that radius. Find the approximate volume of that shell by multiplying the sphere area by the shell thickness.

(d) This shell is thin enough so that we can assume that the electric field energy density remains at the value you found in (b) throughout. Find the total energy of the electric field in that shell by multiplying its energy density by its volume.

(e) Find the mass of the shell's energy, using [1.01].

(f) The mass of a proton is 1.673×10^{-27} kg. Find the fraction (or percentage) of the proton's mass that is the mass of the electric field energy contained in the shell.

Ans.: (a) $E = 1.439 \times 10^{19}$ N/Coul (b) $u = 9.159 \times 10^{26}\,\text{J/m}^3$ (c) $1.257 \times 10^{-42}\,\text{m}^3$
(d) Energy $= 1.151 \times 10^{-15}$ J (e) $m_{Energy} = 1.2789 \times 10^{-32}$ kg
(f) 7.647×10^{-6}, or 0.0007647%.

3.
Nuclear bonds, how much energy?

Nuclear energy comes from the formation or tightening[1] of bonds between nuclear particles (protons and neutrons). That energy is great enough so that its mass equivalent is *measurable* with available laboratory weighing balances.

But, what is it that is to be weighed, exactly?

The main reason for weighing to determine nuclear binding energies is that the reactions themselves are difficult to produce on a small scale. When the question first came up, nuclear reactions had never been observed (except by those looking at the sun or a star, who didn't know that the light they were seeing is powered by nuclear reactions).

We have already alluded to the fact that the answers were known before the question was asked. The answers came not by somehow weighing the energy obtained from an ongoing nuclear reaction, but by accumulating a catalog of the masses of many nuclei before and after their bonds had been formed, or re-arranged.

The prototype of an energy-yielding reaction can be written as follows,

$$\text{mass of reactants} = \text{mass of reacted product} + \text{mass of energy} \qquad [3.01]$$

All particles must be accounted for, or the equation will not be true. The mass of the reactants will include the mass of their bond

---

[1] We use "tightening" to refer not so much to the force between the bonded objects, as to the energy needed to break them apart.

energy; the mass of the reacted products will include the mass of *their* bond energy; the mass of the released energy comes at the expense of the decrease in the energy of the nuclear force fields that occurred in the formation (or tightening) of the bonds.

Since all the particles are still there, they must have become smaller (of smaller mass). Indeed they have. The part of their mass that is electric or nuclear field has diminished. We have seen that in the examples of electric force binding, and, although the nuclear force field is much more complex and cannot be described in a formula, the same conclusion applies. Mass can be lost from that field when nuclear bonds form.

News from the table of isotopes

Isotopes are different species of the same element that differ in the number of neutrons they have in the nucleus. Each isotope is a different particle, and so the table of isotopes, rather than the periodic table, is where we look for the catalog of data.

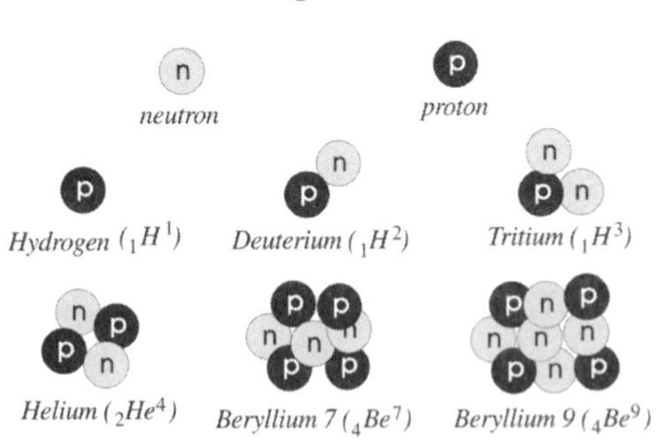

Fig 3.1 Some isotopes

The isotopes of an element are written with two numbers, one a subscript and one a superscript. The subscript is the number of protons in the nucleus; this number is the "atomic number" denoted, "Z," and is the same for all isotopes of the same element. The superscript contains the total of protons and neutrons, and this is called the "atomic mass number," usually given the symbol, "A." The superscript is the number of nucleons in the nucleus.

Example: isotopes of chlorine

The atomic mass of chlorine is listed in the periodic table as 35.453 g/mole. This represents an average, made up of the isotope masses of two isotopes. Both of these isotopes are atoms with 17 protons and 17 electrons, which gives them the chemical characteristics of chlorine. In addition, one of the isotopes, ${}_{17}Cl^{35}$, has 18 neutrons (17+18=35). The other isotope, ${}_{17}Cl^{37}$, has 20 neutrons (17+20=37).

${}_{17}Cl^{35}$ has an isotope mass of 34.9689 g/mole. ${}_{17}Cl^{37}$ has an isotope mass of 36.9659 g/mole. Ordinary chlorine is about 75% ${}_{17}Cl^{35}$ and 25% ${}_{17}Cl^{37}$, giving it its average atomic mass.

Table 3.1 is a table of selected isotopes, which lists their isotope masses, as well as other related quantities. A more complete table is available as an appendix in most physics texts, and exhaustive tables are found in reference books.

One could look for energy-rich nuclear reactions by scanning such a table with possible configurations of reactants before and after in mind. But that is a rather mindless and unsystematic approach. There is a better way.

They all have the same grandparents

All nuclei consist of a number of protons and a number of neutrons. These protons and neutrons were all at one time free and separate particles, bonded to no other particles. For some isotopes the time when they were separate protons and neutrons goes back to shortly after the big bang, but in others the history goes no farther back than the present stars. It doesn't matter.

Every nucleus has a genealogy that goes back to the separate protons and neutrons. Skipping over the intermediate steps, we can write, as an example, the genealogy of chlorine^{35} this way,

$$17\ ({}_{1}p^{1}) \quad + \quad 18\ ({}_{0}n^{1}) \quad \rightarrow \quad {}_{17}Cl^{35} \qquad [3.02]$$

We know very well the characteristics of the ultimate ancestors. All *free* protons have a mass of 1.007825 grams/mole; all *free* neutrons, 1.008665 grams/mole. They are the first entries in the table of isotopes. They have the largest mass per nucleon[2], because, like the free electron, they have the most field energy – there are no bonds yet that have released any of that energy.

Regardless of the sequence through which a nucleus has arrived at its present state of bonding, its present mass represents the original mass of all of its neutrons and protons when they were free, minus the energy that has been lost (released) in the cumulative steps that have built the present configuration.

The difference between the mass of the assembled nucleus and the total mass of the free nucleons out of which it is built, is called the *binding energy* of that nucleus.

Of course, as the number of nucleons in a nucleus increases, the binding energy tends to increase as well. There are more bonds because there are more nucleons. However, if this total binding energy is divided by the number of nucleons, the result, the "average binding energy per nucleon," is a measure of how tightly bonded the nucleus is. The average energy per nucleon does not necessarily increase with the addition of more nucleons.

---

[2] Recall that the term "nucleons" refers to a proton or a neutron.

The binding energy of deuterium

The very simplest nuclear reaction of all is the combination of one neutron and one hydrogen nucleus (a proton) to make one nucleus of deuterium. Deuterium is hydrogen with an added neutron in the nucleus.

Because its nucleus has one proton, the Deuterium atom has just one electron, and is therefore chemically exactly like hydrogen. It is called, "heavy hydrogen," – heavy because the atoms weigh about twice as much as those of ordinary hydrogen. Deuterium is one of the "isotopes" of hydrogen.

The two naturally occurring isotopes of hydrogen are ordinary hydrogen with just the one proton in the nucleus, and deuterium, with 2 nucleons, one a proton and one a neutron. These isotopes are written, ${}_1H^1$, and ${}_1H^2$. For convenience, Deuterium has a special symbol, D, although it is not a different element. It is ${}_1H^2$.

Deuterium makes up about 1/60 of one percent of all natural hydrogen. Chemically it is just like ordinary hydrogen. If one of the hydrogen atoms in a water molecule, H_2O, is deuterium, the molecule is HDO; if both, it is D_2O, which is called "heavy water." It is slightly more dense than ordinary water, but is otherwise indistinguishable. Since 1/60 of one percent of all the ocean water is quite a lot of water, there is no fear of a shortage of heavy water. But separating deuterium-containing water out of ordinary water is a laborious task. Heavy water is not harmful. You can drink it. It tastes no different. You can brush your teeth with it. You can add it in place of ordinary water to your favorite recipe.

Heavy water played an important role in the German atomic bomb project. A raid on a secret heavy water plant, fingered by anti-Nazi patriots in Norway, is said to have irreparably damaged the Germans' chance of finishing an atomic bomb in the Second World War.

Most of the oceans' deuterium was made a long time ago at very high temperatures in the star-matter of the early universe, but deuterium continues to be made in the sun. It requires temperatures of hundreds of millions of degrees, that are difficult to produce in a controlled fashion on earth. The process of deuterium formation in stars actually combines two protons, with the emission of one positive charge as an anti-matter particle called a positron. This is a detail which need not concern us here. We will suppose that the neutron has been produced first, and the deuterium is now made by the joining of a proton and a neutron. The symbol for a proton is ${}_1p^1$ (or ${}_1H^1$) and that for the neutron is ${}_0n^1$.

$$ {}_1p^1 \quad + \quad {}_0n^1 \quad \rightarrow \quad {}_1H^2 \qquad \text{(or, D)} \qquad [3.03] $$

The subscripts add up to the same total on the two sides; that means we have accounted for all the protons. The superscripts add up as well, indicating that the nucleons also are accounted for. (Do not be concerned about electrons. This is a nuclear reaction, and involves only the nuclear particles. The electrons will take care of themselves; in any case they constitute only about 1/2000 of the mass of the atoms, and will balance if the protons do.)

A bond was formed. There is now a single nucleus containing two nucleons. Bond energy was released, and we show that on the right hand side of the reaction,

$$ {}_1p^1 \quad + \quad {}_0n^1 \quad \rightarrow \quad {}_1H^2 \quad + \quad \text{Energy} \qquad [3.04] $$

If we write instead of just their names, the *masses* of the particles in [3.04], including the mass of the energy released, we can write an = sign in the place of the →. This makes it a statement that all the mass that went into the reaction must still be there when the reaction has occurred, if we include the mass that went off with the released energy.

$$\text{mass}({}_1p^1) + \text{mass}({}_0n^1) = \text{mass}({}_1H^2) + \text{mass(Energy)} \quad [3.05]$$

From Eq 3.05 we can determine the energy that is released in the formation of deuterium from the separate protons and neutrons.

It requires only that we can measure the mass of a standard number of each of the particles, and that is something that had been done. The chemists who explored the periodic table had long been very meticulously measuring the atomic masses of the elements. We require now the mass of each isotope separately, but that is just more of the same task.

Isotope tables give the mass of the isotopes in grams per mole. A "mole" is a number, like a "dozen." That number is about 6.02×10^{23}, called Avogadro's number, written, N. This number was originally the number of atoms of the lightest element (hydrogen) that make up one gram of hydrogen. The mole has now been re-defined as the number of atoms of the ${}_6C^{12}$ isotope in 12.00000000 grams of ${}_6C^{12}$.

===

Calculation of the binding energy of deuterium

The Table of Isotopes (Table 3.1) gives us:

Mass of one neutron:	1.008665 g/N
Mass of one proton (${}_1H^1$)	1.007825 g/N
Mass of 1 nucleus of (${}_1H^2$)	2.014102 g/N

Substituting these numbers in [3.05],

$$1.008665\,\text{g}/N + 1.007825\,\text{g}/N = 2.014102\,\text{g}/N + \text{mass(energy)} \quad [3.06]$$

which gives mass(energy) = 0.002388 g/N

This quantity is the mass of the energy released in the reaction when one mole of neutrons and one mole of protons bond to produce one mole of deuterium nuclei. (The symbol, N, is used to mean "mole.")

To express $0.002388\,g/N$ in energy units, Joules per mole, it is necessary first to express the mass in kg, because $1\,J = 1\,kg\,m^2/s^2$.

$$\text{mass(energy)} = .002388\,g/N \times (1\,kg/1000\,g) = 0.000002388\,kg/N$$

Using $E = mc^2$, with $c = 3 \times 10^8 m/s$,

$$E = (0.000002388\,kg/N) \times (3 \times 10^8\,m/s)^2 = 2.145 \times 10^{11}\,J/N$$

===

It is interesting to see what this result means in every-day terms. Our electricity consumption is listed on electric bills in KiloWatt Hours.

One Watt is one Joule/sec. It is a rate of energy consumption. A 60 Watt light bulb uses 60 Joules of electric energy per second. A Kilowatt-hour, which is the unit of billing of energy on most electric bills, is 1000 Watts for one hour, or 3,600,000 Joules.

The binding energy released when one mole (2 grams) of deuterium is made out of free protons and free neutrons is

$$2.145 \times 10^{11}\,J \times (1\,\text{KW-hr} / 3{,}600{,}000\,J)$$

which is about 60,000 KW-hr. This is enough to supply an average home with electrical energy for about 10 years.

It is not practical to make deuterium out of protons and neutrons here on earth. In general, we can't build nuclei out of free protons and free neutrons. We will now develop a strategy to find reactions that might be made to go here on earth, hopeful that this will cut us at least a share of this enormous resource of energy.

The "binding energy per nucleon" graph

To find a nuclear reaction that will give us energy, we must search for nuclei whose nucleons can be re-arranged in such a way that they end up more tightly bonded than before. There is a systematic way to search for such possibilities.

"More tightly bonded" means that the *average binding energy per nucleon* has become greater. This is the quantity that we need in our catalog.

It means that in our catalog under deuterium we want not its total binding energy, 2.145×10^{14} J/mole, but half that, because there are two nucleons. The average binding energy *per nucleon* of deuterium is 1.073×10^{14} J/mole.

The search for an energy-yielding nuclear reaction begins with a graph of binding energies/nucleon. Recall that a tighter bond is one from which more energy *has already been released*, so that less energy is left. Our interest is in how much energy is still there, not how much has already been removed, so for us the most instructive graph is one that tells how much farther down in energy this nucleus can go.

The most tightly bonded nucleus known is that of iron, ${}_{26}Fe^{56}$. It is a useful convention to refer to the binding energy remaining in this isotope as the zero of the graph. The binding energy of iron is 1.41552×10^{-12} Joules per nucleon. This is a small number because it is per nucleon, not per mole of nucleons. To avoid dealing with numbers with such large negative powers of ten, the unit of energy used conventionally in this field is the MeV, or "million electron volts." The conversion is

$$1 \text{ MeV} = 1.6 \times 10^{-13} \text{Joules}$$

In these units, the binding energy of iron is 8.847 MeV/nucleon.

Table 3.1 Isotope Table

Isotope name	no of protons Z	no of neutrons	atomic mass no A	Isotope mass (g/mole)	Binding energy (g/mole)	Bdg En / nucleon (g/mole)	remaining B.E. (MeV)/nucleon)
neutron $_0n^1$	0	1	1	1.008665	0	0	8.847
proton $_1p^1$ ($_1H^1$)	1	0	1	1.007825	0	0	8.847
deuterium $_1H^2$	1	1	2	2.014102	0.002388	0.001194	7.727625
tritium $_1H^3$	1	2	3	3.016049	0.009106	0.003035	6.001375
helium $_2He^4$	2	2	4	4.002603	0.030377	0.007594	1.72739
lithium $_3Li^6$	3	3	6	6.015123	0.034347	0.0057245	3.48028
beryllium $_4Be^9$	4	5	9	9.012183	0.062442	0.006938	2.34263
boron $_5B^6$	5	6	11	11.009305	0.08181	0.007437	1.87456
carbon $_6C^{12}$	6	6	12	12.000000	0.09894	0.008245	1.11731
nitrogen $_7N^{14}$	7	7	14	14.003074	0.112356	0.008025	1.32316
oxygen $_8O^{16}$	8	8	16	15.994915	0.137005	0.008563	0.81936
neon $_{10}Ne^{20}$	10	10	20	19.992439	0.172461	0.008623	0.76289
aluminum $_{13}Al^{27}$	13	14	27	26.981541	0.241494	0.008944	0.46179
sulfur $_{16}S^{32}$	16	16	32	31.972072	0.291768	0.009118	0.29911
potassium $_{19}K^{39}$	19	20	39	38.963708	0.358267	0.009186	0.23481
titanium $_{22}Ti^{48}$	22	26	48	47.947947	0.449493	0.009364	0.06784
manganese $_{25}Mn^{55}$	25	30	55	54.938046	0.517529	0.009410	0.02548
iron $_{26}Fe^{56}$	26	30	56	55.934939	0.528461	0.009437	0
cobalt $_{27}Co^{59}$	27	32	59	58.933198	0.555357	0.009413	0.02247
copper $_{29}Cu^{63}$	29	34	63	62.929599	0.591936	0.009358	0.03843
bromine $_{35}Br^{79}$	35	44	79	78.918336	0.736799	0.009327	0.10334

Table 3.1 Isotope Table (continued)

Isotope name	no of protons Z	no of neutrons	atomic mass no A	Isotope mass (g/mole)	Binding energy (g/mole)	Bdg En / nucleon (g/mole)	remaining B.E. (MeV)/nucleon)
krypton $_{36}Kr^{84}$	36	48	84	83.911506	0.786114	0.009359	0.07340
krypton $_{36}Kr^{94}$	36	58	94	93.915	0.86927	0.009248	0.17742
molybdenum$_{42}Mo^{98}$	42	56	98	97.905405	0.908485	0.009270	0.15613
palladium $_{46}Pd^{106}$	46	60	106	105.90348	0.97637	0.009211	0.21165
silver $_{47}Ag^{107}$	47	60	107	106.905095	0.98258	0.009183	0.23795
tin $_{50}Sn^{120}$	50	70	120	119.902199	1.095601	0.009130	0.28762
iodine $_{53}I^{127}$	53	74	127	126.904477	1.151458	0.009067	0.34706
barium $_{56}Ba^{138}$	56	82	138	137.90524	1.24349	0.009011	0.39938
barium $_{56}Ba^{139}$	56	83	139	138.905	1.252395	0.009010	0.40009
tungsten $_{74}W^{184}$	74	110	184	183.95095	1.58125	0.008594	0.79036
gold $_{79}Au^{197}$	79	118	197	196.96656	1.674085	0.008498	0.88022
lead $_{82}Pb^{204}$	82	122	204	203.973044	1.725736	0.008459	0.91623
lead $_{82}Pb^{206}$	82	124	206	205.97446	1.74165	0.008455	0.92080
lead $_{82}Pb^{208}$	82	126	208	207.97664	1.7568	0.008446	0.92873
lead $_{82}Pb^{210}$	82	128	210	209.98418	1.76659	0.008412	0.96043
lead $_{82}Pb^{214}$	82	132	214	213.9998	1.78563	0.008244	1.02444
radium $_{88}Ra^{226}$	88	138	226	226.025406	1.858964	0.008226	1.13559
uranium $_{92}U^{235}$	92	143	235	235.043925	1.91507	0.008149	1.20709
uranium $_{92}U^{238}$	92	146	238	238.050786	1.934204	0.008127	1.22802
plutonium $_{94}Pu^{239}$	94	145	239	239.052158	1.939817	0.008116	1.23789

Fig 3.2 "Binding Energy Remaining" (relative to zero at Iron)

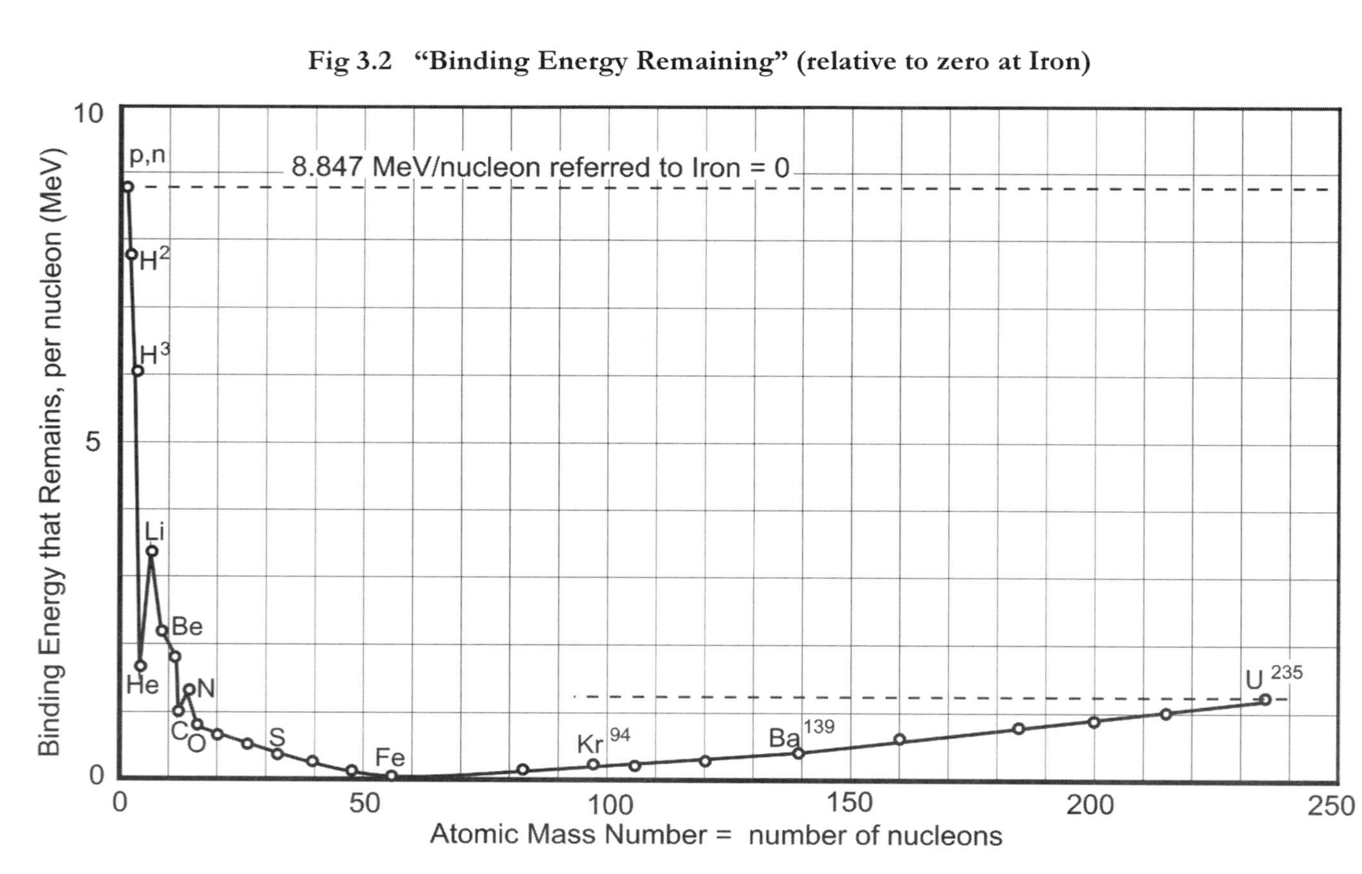

If separate protons and nucleons were packaged as iron nuclei, they would yield up 8.847 MeV/nucleon in binding energy. This is the amount of energy that each separate and unbound proton or neutron has in the form of "binding energy remaining" that it could yield up if it were packed with other nucleons in the tightest possible way, in an iron nucleus.

For any other nucleus, the "binding energy remaining" is the binding energy it had as separate protons and neutrons, minus the binding energy it has already given away in its own assembly process. The difference between 8.847 MeV/nucleon and its "binding energy" is what it has left to give away if its nucleons could be re-packaged as iron.

The graph of Fig 3.2 is a graph of "binding energy remaining, per nucleon," graphed against the isotopes, with the numbers on the horizontal axis being the atomic mass numbers, A, the numbers of nucleons in the isotope. The numbers graphed in Fig 3.2 are shown in the final column of Table 3.1.

The graph of Fig 3.2 shows that binding energies are not a helter-skelter assortment, but follow a definite pattern.

==

Example 1: Calculation of Binding energy of Tritium

A "binding energy" problem can be recognized by the fact that the problem does not say anything about starting reactants. That is because, in a binding energy calculation, by definition, the reactants are the separate protons and neutrons that have gone into making a particular isotope.

Find the binding energy of tritium, $_1H^3$

1. Read the subscript and superscript to find the number of protons and neutrons in the nucleus. The number of protons is given in the subscript.
   In $_1H^3$ there is one proton (it is an isotope of hydrogen).

The superscript is the total of protons and neutrons, so the number of neutrons is the difference between the superscript and the subscript.
In $_1H^3$ there are two neutrons (3 minus 1)

2. Write the reaction, with only separate protons and neutrons as the starting material, on the left side of the reaction,

$$1\,({}_1p^1) \;+\; 2\,({}_0n^1) \;\rightarrow\; {}_1H^3 \;+\; \text{energy}$$

3. Find the total mass of the protons and neutrons that went into making $_1H^3$.
   mass of 2 neutrons = 2×1.008665 g/mole = 2.017330 g/mole
   mass of 1 proton = 1×1.007825 g/mole = 1.007825 g/mole
   total 3.025155 g/mole

4. Subtract the isotope mass of $_1H^3$(look it up) 3.016049 g/mole
   find the difference 0.009106 g/mole

This is the mass defect, the total binding energy per mole

5. To express this in energy units (Joules), use Eq [1.01] $E = mc^2$

$$E = (0.009106\,\text{g/mole}) \times (1\,\text{kg}/1000\,\text{g}) \times (3 \times 10^8\,\text{m/s})^2$$
$$= 8.1954 \times 10^{11}\ \text{Joules/mole}$$

(conversion of grams to kg is necessary to give Joules)

6. To find binding energy per nucleon, divide by the number of nucleons in the nucleus (3) and divide by Avogadro's number, N

$$E = (8.1954 \times 10^{11}\ \text{Joules}/N)/3\ \text{nucleons}/(6 \times 10^{23}\ \text{nuclei}/N)$$
$$E = 4.553 \times 10^{-13}\ \text{Joules/nucleon (average)}$$

7. To express this energy in MeV, use $1\ \text{MeV} = 1.6 \times 10^{-13}\ \text{Joule}$

$$E = 4.553 \times 10^{-13}\ \text{Joules/nucleon} \times 1\ \text{MeV}/1.6 \times 10^{-13}\ \text{Joule}$$
$$= 2.846\ \text{MeV/nucleon}$$

8. To find "Binding energy *remaining*," based on binding energy of iron as the zero: Subtract binding energy of $_1H^3$ from the binding energy of iron:
   8.847 MeV/nucleon – 2.846 MeV/nucleon = 6.001 MeV/nucleon

   "Binding energy remaining" of $_1H^3$ is 6.001 MeV / nucleon

===

The example tells us that there is 6.001 MeV of binding energy remaining, per nucleon, in tritium. That is what is available, if the nucleons could be re-packaged as tightly as the nucleons of the iron nucleus, which is the most tightly packaged nucleus we know. More than that can not be got. The question is, how great a share of that can we, in practice, obtain?

The graph as a road map to the bomb

We can not take free protons and free neutrons and let them form bonds and release that binding energy to us. The next best shot at getting energy is to find some already existing nuclei in which the nucleons, for whatever reason, are not packed as tightly as they might be. Can we re-arrange these same nucleons in a way that allows their bonds to each other to tighten?

That means getting them to form other isotopes whose "remaining binding energy per nucleon" is less than it is in the existing nuclei, meaning they are lower down in the graph.

It is clear how the graph of Fig 3.2 can help with this. That graph tells us exactly which are the tightly packed nuclei and which are the less tightly packed. Our intuition about how to get energy from a rolling ball is quite appropriate here. The higher the ball, the less tightly bonded the ball is to earth (by gravity). This is why a ball will increase in speed as it rolls *down*, gaining kinetic energy. Exactly so with these nuclei. If those that are higher up in the graph can be made to roll down the graph, their nucleons will pack more tightly, and release energy.

The graph reveals where the rearrangements are that would be favorable for the release of energy. Those isotopes whose points are higher up have more binding energy remaining, those lower have less. If they can be made to undergo a rearrangement from a higher to a lower place on the graph, we might be able to harvest some of the binding energy they lose.

Why the graph has the shape that it does

The graph of Fig 3.2 shows remaining binding energy per nucleon for the entire gamut of elements, graphed from left to right as a function of the isotope mass number, A (the number of nucleons in the nucleus).

It is not surprising that there is a pattern – that these values are not random. What perhaps is surprising is what that pattern looks like. Why do the binding energy values first fall, then rise?

Let us start at the left. When A is 1, it means we have a single neutron or a single proton, no binding energy has yet been spent, so the "binding energy remaining" is highest. Moving right, to $A = 2$, we have the deuterium nucleus. Because of the fairly large amount of energy that was released in the forming of this bond, there is considerably less energy left, per nucleon, than when there were no bonds, at $A = 1$. The curve shoots downward.

If we add another neutron to deuterium we get tritium, ${}_1H^3$, another hydrogen isotope. Tritium does not occur naturally because it is unstable, and dissociates (is radioactive) spontaneously with a half life of about 12 years. But while it lasts, it is a nucleus which has formed *two* new bonds, one bond between the new neutron and each of the nucleons of deuterium, and so has released more energy, and, of course, more mass. Down swoops the curve some more.

Details aside (there are considerations of stability beyond just a count of bonds), at the beginning of the curve, there is a relatively steep decline, since with the addition of each new nucleon (proton or neutron) the number of bonds increases by the number of nucleons already there, since the new nucleon bonds to *all* the nucleons already present.

This goes on quite consistently, except for a spike at $A = 4$, which is helium. The helium nucleus is the well known "alpha" particle, which is an extraordinarily stable and tightly packed arrangement.

Yet, it does not continue this decline forever. After about $A = 8$, beryllium, the "binding energy remaining" per nucleon declines less and less rapidly. This occurs for a combination of two reasons. The first is a factor called, "saturation," which limits the number of strong force bonds that a nucleon can have to bonds with near neighbors.

The other factor is that the nuclear strong force "glue" is in competition with a very vigorous electrical force of repulsion between every proton and every other proton, tending to blast the whole nucleus back apart into its separate protons and neutrons. At short distances, the electric force of repulsion is no match for the nuclear force of attraction. But the electric force does not diminish with distance as rapidly as the nuclear strong force. So, as distances increase, the strong force finds it harder and harder to maintain, never mind increase, the average net glueing force.

Around $A = 50$ to 60, in the region of the periodic table around iron and nickel, the contest is a standoff. As more nucleons are added, the average nuclear attraction increases at about the same rate that the electric repulsion increases, and the curve bottoms out. After that, with distances still increasing, and more nucleons being added, the proton repulsion effect grows more rapidly than the nuclear attraction effect, and the nuclei actually are less well glued together.

The nuclei with values of A higher than 60, to the right of the bottom plateau on the graph, are increasingly less stable; they are at a higher energy level and would actually prefer to drop back down to the lowest energies around iron.

How these less stable nuclei came about at the time that these aggregations were assembled in the stars is a reasonable question. Nuclei formed randomly. Even some nuclei that were not in their lowest energy state, were kept from reverting to lower energy configurations by an "energy barrier," like a rim around a depression at the top of a hill. They stay there because in trying to leave the depression, they would first have to go *up* to "get out." Once a nucleus of, let us say, gold, with $A = 197$, is formed, it would require the arrangement to surmount an energy "barrier," a configuration still higher in energy than that of gold, from which to roll down to the energy level of iron.

We will see that this energy barrier, both because it is there and also because it is not insurmountable, is of great importance to our ability to use fission of uranium as a source of energy.

Search for a nuclear fuel

The graph of "binding energy remaining" as a function of atomic size is exactly what the early nuclear designers needed to determine where to look for nuclear fuels, as well as where not to bother.

It is clear where on the graph the nuclei are that are not yet bonded as tightly as their constituent nucleons could be. Certainly iron and its neighbors on the periodic table are *not* among these, for these nuclei are already as low on the graph as they can be. The ones on the extremes at the left and the right are the candidates.

But the graph can also serve as a chart to search among those candidates, to find ones that can actually be induced to rearrange their nucleons under earth conditions. These will be potential nuclear fuels. Almost all nuclear rearrangements are possible at temperatures of several hundred billion degrees, temperatures that prevailed in the early universe, and that occur still in large stars near the end of their life cycle, just before they explode as super nova.

But on earth, the inducements are limited.

The first nuclear engineers were trying to build a bomb. The physicists had laid the theoretical groundwork in the 1930's, so it was clear at the outbreak of World War II in 1939 that the war may be won by the first side to be able to build a nuclear bomb. Calculations such as the ones we have already done in the last two chapters, showed that a weapon the size of a large torpedo could conceivably release enough energy to destroy a city. Neither side wished to be in a war in which the other side could wreak such destruction. And so, both sides worked feverishly on any possible way to make this energy source into a weapon of war.

Many books have been written about the Nazi German nuclear bomb effort. Some top German physicists worked on the project. It is not clear whether they really wanted the project to succeed, or only pretended to be working towards that end. A memo from the team to the Nazi high command rather late in the war suggests that they had over-estimated by far the amount of nuclear material that would be needed, to the point where it appeared impossible to amass a quantity sufficient for a bomb. Whether this over-estimate was an intentional mistake, or an actual one, is not certain. It seems to have stalled the German effort. Meanwhile, the United States, with its larger resources of space, industrial resources, and scientific personnel, succeeded in producing two bombs. Germany had already been defeated by this time, but two Japanese cities felt the awsome destruction of these bombs, which resulted in Japan's surrender.

How to make the nucleons tighten their bonds

From the graphs, it was clear that the process of bond forming or bond tightening could come either of two ways:

(1) from nuclei at the left end being assembled into larger nuclei, a process called nuclear *fusion* (meaning "coming together") to move the nucleons into an arrangement that is down and to the right on the graph, or

(2) from nuclei on the right being rearranged to move their nucleons down and toward the left in the graph, where bonds were on the average tighter, and energy would be released. Such a rearrangement would require making smaller units out of a larger nucleus, called nuclear *fission* (breaking apart). On the right side of the graph, to the right of iron, the smaller units are more tightly packed, as evidenced by their lower position on the graph.

The possibility of imitating what was known to occur in the sun at its high temperatures, combining two deuteriums or a deuterium and a tritium, was considered. The conditions required seemed hard to achieve, and the attack on the problem was shifted to the other end of the graph.

As one moves to the higher end of the periodic table, the nuclear glue is less and less able to hold the nucleus together against the repulsions of the protons. It is the reason why no element higher in atomic number than that of uranium, 92, occurs naturally. All elements higher in the periodic table are unstable and dissociate spontaneously, and hence aren't around anymore. Uranium seemed to be the element on the boundary, teetering between survival as a nucleus and instability leading to radioactive dissociation.

Uranium was looked upon as the element most likely to be susceptible to a nudge that would cause it to divest itself of excess baggage. That it would split, rather than get rid of a small radioactive fragment was a surprise. Uranium fission was first observed in a laboratory in 1939, somewhat by accident.

Problems

3.1 Find the BINDING ENERGY of the common isotope of Beryllium, $_4Be^9$. Express it in

(A) grams per mole (*N* atoms) of beryllium; and in

(B) Joules per mole (*N* atoms) of beryllium; [Reminder: the units of the Joule are **kg** m^2/sec^2] and in

(C) Joules per nucleon of beryllium

(D) Find the "Binding energy remaining" in MeV / nucleon [check answers to (A) and (D) with Table 3.1]

3.2 The isotope mass of Magnesium ($_{12}Mg^{24}$) is 23.9850 g/*N*. Find the binding energy of one mole of Magnesium. This is a two-step problem: First find the mass defect (per mole). Then express the binding energy in Joules.

3.3 The isotope mass of Neon-22 ($_{10}Ne^{22}$) is 21.9914 g/*N*. Find the binding energy of one mole of Neon-22 (not the common Neon isotope). This is a two-step problem: First find the mass defect (per mole). Then express the binding energy in Joules.

4. Energy from repackaging nucleons

In chapter 3 we wrote, in the manner in which chemical reactions are written, the nuclear reaction in which a proton and a neutron combine to form a deuterium nucleus [3.03]. In Eq 3.04 we added for completeness a term for the energy (or mass). By substituting for each constituent the value of its mass (from Table 3.1), we solved the equation to find how much mass was released.

Binding energy a special case of energy yield from a nuclear reaction

When we assemble separate protons and neutrons to build a nucleus, we obtain the *binding energy* for the nucleus, the energy that is released, cumulatively, by the totality of steps in assembling the nucleus out of separate protons and neutrons.

In Example 1 of chapter 3, the binding energy of Tritium ($_1H^3$) was calculated. At the end of step 5 of that sample calculation, one obtains the *energy yield* in Joules/mole of such an assembly process.

We will now be looking at the calculation of binding energy as a special case of a more general problem, in which the starting materials are not necessarily separate protons and neutrons, but may be one or more nuclei (and other nucleons), which react to form one or more resulting nuclei (and other nucleons). In this broader category of nuclear reactions might be one that can actually be induced to happen, whereas the assembly from separate protons and neutrons is not within our power to perform.

To review the method, and to establish a form for solving the more general reactions, we review steps 2 to 5 of Example 1 in a new example below showing the assembly of Helium from two protons and two neutrons. The determination of "binding energy remaining" in MeV/nucleon is not of interest here. We stop with step 5, having found the energy yield per mole of reactants.

Free and separate nucleons to helium

$$2\,({}_1p^1) \quad + \quad 2\,({}_0n^1) \quad \rightarrow \quad {}_2He^4 \quad + \text{energy} \qquad [4.01]$$

and from the isotope table,

$$2\,(1.007825\,g/N) + 2\,(1.008665\,g/N) = 4.002603\,g/N + \text{energy}$$

$$4.032980\,g/N = 4.002603\,g/N + \text{energy}$$

$$\text{Energy Yield} = 0.030377\,g/N \qquad [4.02]$$

To express the Energy Yield (E.Y.) in energy units, one uses the mass-energy equivalence equation, [1.01], $E = mc^2$, where "m" must be given in kg so that E.Y. will be in Joules. 1 kg = 1000 g.

$$\text{E.Y.} = (0.030377\,g/N) \times (1\,kg/1000g) \times (3 \times 10^8\,m/s)^2$$

$$\text{Energy Yield} = 2.73 \times 10^{12}\,\text{Joules}/N \qquad [4.03]$$

This is the *binding energy* for one mole of helium. We now look upon it as the *Energy Yield* of the reaction [4.01]. It is called the *binding energy* only when the reactants (the starting materials, on the left side of the reaction equation) are separate protons and neutrons.

Reactions of rearrangement

The graph of Fig 3.2, showing "binding energy remaining," steers us to nuclear *re-assembly* reactions. These begin with already assembled nuclei that perhaps could be re-assembled differently to release more binding energy. If we can make any such rearrangements happen, we might gain access to the additional binding energy that would be released. Any rearrangements that move downward in the graph of Fig 3.2 would accomplish this end. The farther down they move, the greater will be the energy yield.

The search for such re-assembly that can in reality be made to go is the search for nuclear fuels. There are reasons why this search is difficult. They have to do with activation – finding out how to make it happen, and then being able to do what has to be done to make it happen, and happen safely. These questions are the subject of Part II of this book.

In this chapter we will go through two examples in which we calculate the yield in energy from nuclear re-arrangement reactions.

===

Example 1: Helium to carbon

First a reaction that can not be made to happen on earth, but whose simple and dramatic arithmetic makes it an ideal first example.

A nucleus of the common isotope of carbon contains six protons and six neutrons, and is therefore called ${}_{6}C^{12}$. You will notice that ${}_{6}C^{12}$ has exactly three times as many of everything as does ${}_{2}He^{4}$. Thus, although we can not make this reaction occur, it is possible in theory to take three Helium atoms and combine them to make one Carbon atom, with nothing added and nothing left over.

The assembly of three heliums into one carbon did, in fact, occur in the early universe, and does still today in some stars, but it requires very high temperatures and takes times measured at least in millions of years to occur in significant quantities. It is, nevertheless, one of the ways in which carbon came to exist in the universe.

Naively, ignoring what we know about the mass of bond energy, one would expect the nucleus of ${}_6C^{12}$, since it has exactly three times as many of everything as a nucleus of ${}_2He^4$, to have exactly three times the mass of one nucleus of ${}_2He^4$. In other words, the isotope mass of ${}_6C^{12}$ ought to be three times the isotope mass of ${}_2He^4$. Either way, there are 6 protons and 6 neutrons.

When the isotope masses were measured, that of ${}_2He^4$ was found to be 4.002603 g/mole, and that of ${}_6C^{12}$ was found to be 12.000000 g/mole. (The isotope mass of Carbon is an integer number because ${}_6C^{12}$ has become the standard for the definition of Avogadro's number.)

The mass of three moles of ${}_2He^4$ is $3 \times (4.002603\text{ g})$ or 12.007809g.

If we re-package those very same $6N$ neutrons and $6N$ protons in one mole of ${}_6C^{12}$, their mass is only 12.000000g!

What happened to the missing 0.007809g? Of course you know the answer: it was released. It is the "energy yield" of this nuclear reaction,

$$3\ {}_2He^4 \rightarrow {}_6C^{12} + \text{Energy yield} \qquad [4.04]$$

substituting masses,

$$3 \times (4.002603\text{ g}/N) = 12.000000\text{g}/N + \text{Energy yield} \qquad [4.05]$$

$$\text{Energy yield} = 12.007809\text{ g}/N - 12.000000\text{ g}/N$$

$$\text{Energy yield} = 0.007809\text{ g}/N \qquad [4.06]$$

$$= (0.007809\text{ g/N}) \times (1\text{ kg}/1000\text{ g}) \times (3 \times 10^8\text{m/s})^2 \qquad [4.07]$$

$$\text{Energy yield} = 7.028\times10^{11}\text{Joules/mole} \qquad [4.08]$$

"Per mole" here means per 3 moles of helium or per mole of carbon produced.

==

The activation of the helium-to-carbon reaction discussed above is very difficult to achieve, principally because it involves getting three nuclei, each positively charged with two protons, to come close enough to each other for the strong nuclear force to engage. The repulsion forces of the positively charged nuclei is too much to overcome. Only at extremely high temperatures is the kinetic energy of the nuclei ever great enough to be able to overcome the repulsion that wants to slow them down and turn them around.

Deuterium and Tritium to Helium

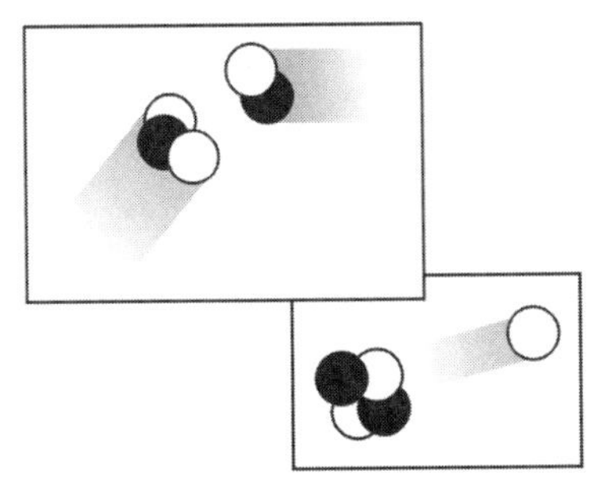

Fig 4.1 Tritium nucleus (top, left) is joined by a deuterium nucleus. It bonds tightly with the new proton, forming helium, kicks out a neutron.

A reaction that can be made to take place, because its activation requirement is less prohibitive, is the merging of two hydrogen nuclei, one a nucleus of deuterium (${}_1H^2$), and the other a nucleus of tritium (${}_1H^3$). This requires temperatures of several hundred million degrees, and while this is not easy to achieve in controlled situations, the use of a uranium bomb-within-a-bomb to create such a temperature makes the so-called thermonuclear fusion reaction described here possible under bomb conditions. However, fifty years of effort has not produced a practical way to activate it under reactor conditions in which it could generate energy for industry and home. Some day, it is hoped, this will be possible.

The reaction, which is called " thermo-nuclear" fusion, produces helium, and a neutron is left over as well. The energy is released partly as kinetic energy of the neutron.

Example 2: Thermo-nuclear fusion

The reaction is,

$$ {}_1H^2 + {}_1H^3 \rightarrow {}_2He^4 + {}_0n^1 + \text{Energy yield} \quad [4.09]$$

The isotope table gives the masses,

$$2.014102\,g/N + 3.016049\,g/N = 4.002603\,g/N + 1.008665\,g/N + \text{Energy yield} \quad [4.10]$$

$$\text{Energy yield} = 0.0189\,g/N \quad [4.11]$$

$$\text{Energy yield} = (0.0189\,g/N) \times (1\,kg/1000\,g) \times (3 \times 10^8 m/s)^2$$
$$= 1.701 \times 10^{12}\,\text{Joules/mole} \quad [4.12]$$

Problems

4.1 Find the energy yield in grams per mole and in Joules per mole for the following nuclear reaction:

$_1H^2 \quad + \quad _7N^{14} \quad \rightarrow \quad _6C^{12} \quad + \quad _2He^4 \quad + \quad \text{Energy Yield}$

(Use Table 3.1 for isotope masses)

4.2 Find the energy yield in grams per mole and in Joules per mole for the following nuclear reaction:

$_2He^4 \quad + \quad _7N^{14} \quad \rightarrow \quad _8O^{17} \quad + \quad _1H^1 \quad + \quad \text{Energy Yield}$

(Use Table 3.1 for isotope masses.

The isotope mass of $_8O^{17}$ is 16.9968 g/N)

4.3 Find the energy yield in grams per mole and in Joules per mole for the following nuclear reaction:

$_0n^1 \quad + \quad _{92}U^{238} \quad \rightarrow \quad _{92}U^{239} \quad + \quad \text{Energy Yield}$

(Use Table 3.1 for isotope masses

The isotope mass of $_{92}U^{239}$ is 239.0543 g/N)

Answers:

4.1 E.Y. = 1.314×10^{12} J/N; 4.2 E.Y. = 9.9×10^{10} J/N;

4.3 E.Y. = 4.68×10^{11} J/N;

PART II

The Technology of Nuclear Energy

5. Activation

Summary of Part I

In part I we found that the essential characteristic of anything that provides us with energy, is that it consist of two or more components that attract each other and would, if allowed, form a bond (or several bonds), but have not yet done so: gasoline and oxygen, an electron and a proton, the earth and water at a high level, a loosely held electron and an oxidized state of a molecule in a battery, a proton and a neutron, deuterium and tritium. In the process of coming together under the force that creates the bond, Work is done, and energy is released.

To search for fuels, we must look for situations in which a bond of some sort can be formed (or tightened or replaced by a stronger one). As that bond is formed, energy is released. The energy is not *in the bond.* The bond energy is there *before* the bond is formed; once the bond is formed, the energy has been released and is no longer there. Energy is in the fuel before the bond has been formed.

We found that all this applies even to nuclear energy; that nuclear energy is available from the nuclear bond forming or tightening, and would be available even in the absence of the mass–energy equivalence, $E = mc^2$.

We found that mass-energy equivalence means that a release of energy is *always* accompanied by a release of mass, whether the process is chemical, electrical, or nuclear. We noted that ordinarily the quantity of energy available is most easily found by measuring the energy directly, as in heat released by burning coal, but that in a nuclear reaction it is most easily measured by *weighing* the mass difference in the fuel before and after.

There is yet one more qualification that a fuel must meet. It must require activation. Not "may," – "must." Or it isn't a fuel.

What is activation? It is whatever gets the process of making (or strengthening or replacing) the bond happen. It is the spark plug that ignites the gas-air mixture, the match that lights the log, the switch that opens the way for electrons to pass from one side of the battery to the other.

Think of activation as lifting the ball over the rim of the depression in the top of the mound, after which the ball can roll down.

Activation, indeed, may appear to be just one more nuisance for the engineer. One more hurdle to overcome. But look at it a different way.

What if there were no depression in the top of the mound, with a rim around it? What if the ball just sat at the top of the mound?

Exactly right! The ball would not stay there, but would roll down. When you come along to find balls to roll down as a source of energy, you would find they had all rolled down, and were at their lowest level already. There would be no energy left for you to extract from this fuel.

What if gasoline and air did not require a spark to set them off, but combined without any activation? Gasoline would burn all by itself, and it would have done so long ago. Even if you could make fresh gasoline, you could not keep it in the gas tank, because it would burn there as well. Fortunately for us, gasoline molecules and oxygen molecules, at ordinary temperatures, repel when they come near each other. Like the ball that has to be lifted out of the depression and over the rim before it rolls down, gas and oxygen have to receive a nudge to push them over the range of repulsion (which is appropriately called an "energy barrier") to the point where attraction of the chemical bond takes over, forms the strong bond, gives up the energy of combustion.

The switch that holds the electrons in their high energy level in the zinc of the battery also constitutes an energy barrier, holding the electrons in their high energy atomic orbitals, ensuring that the battery has not discharged before we needed it.

The requirement of activation is what keeps all our energy sources from yielding up their energy to the great outdoors before we are ready to use it. It gives *us* the control over the moment when the energy is yielded up; so that only when we are ready to use it does the fuel "burn."

The world was once full of fuels that required no activation. These fuels formed their bonds long ago, and are no longer "fuels," no longer have any energy to give. Iron and oxygen require no activation, although they combine slowly. But iron rusts, and iron left for a few millennia becomes iron ore, which is why we find iron as iron oxide in the ground. To make iron out of it, we have to drive the oxygen off, by supplying energy in a blast furnace.

All the oxygen and hydrogen have long since made oceans full of water. This is why there is very little naturally occurring hydrogen in our atmosphere. (Although, as chemistry students know, even hydrogen needs to be lit to burn with oxygen.)

So it is with our nuclear fuels. We therefore study the activation process in nuclear fuel with humble appreciation for the role that activation plays in preserving that fuel for us, so that we can have its energy at a time of our choosing.

Because we do not have the luxury of being allowed to design the activation process that unlocks nuclear energy release, some of the most difficult problems are associated with activation.

Our most yearned-for nuclear fuel, deuterium and tritium, has such a high activation requirement that we have not yet figured out how to activate this fuel without setting it off in the fireball of a thermonuclear bomb.

Uranium fission

In part I we determined that there were essentially two categories of nuclear reaction that yield energy, those reactions in which two small nuclei join, producing new bonds between them, and those in which large nuclei shed some of their nucleons, and form smaller, more tightly bonded, isotopes.

In both cases, the rearrangements that occur result in a repackaging of the nucleons moving them lower in the "binding energy remaining" graph, meaning that energy is released.

Uranium reactors and weapons fall into the second of these categories. In those reactions, the uranium nucleus splits into smaller nuclei, such as those of barium and krypton.

But energy is not released in the breaking of bonds. To break bonds apart always requires that energy be supplied. (It always *requires* energy to stretch a spring, to separate two magnets that are stuck together, to lift a weight that has fallen to earth, to drive the oxygen off the iron oxide.) It is in the nature of a bond that it holds things together. Therefore one has to invest energy from the outside to stretch the bond or to break it.

Energy has to be supplied to split the uranium nucleus into the two smaller fragments. Fortunately so. If that were not so, all the uranium would already have split, and it would not be a fuel. The energy to split uranium is the activation energy that we have to supply to break the bond that holds the nucleons together in the uranium nucleus.

It is only *after* the uranium nucleus has split that energy is released. The smaller fragments of the "fission" are more tightly bonded. It is as though, once split, the fragments each make a tighter fist, the nucleons draw to each other in the fragments more strongly because they are closer together. How do we know? Because the energy content per nucleon is less – because the fragment nuclei are lower on the "binding energy remaining" graph.

The discovery of fission

How did scientists discover that this could be accomplished? Somewhat by chance, as is often true in science, but also by looking in the right places.

The craft of the particle physicists is to hit particles with other particles. Not one at a time, of course; they can't be aimed that precisely. But a stream of one type of particle is targeted to collide with a stream of other particles. These are called "particle beams."

It was a natural thought that, because uranium teeters at the limit of nuclear size that is stable, it might be nudged into seeking to attain a more comfortable (lower energy) configuration by giving it a kick, that is, by hitting it with some energetic little particle.

James Chadwick had discovered neutrons, and methods had been developed for producing streams of neutrons. Neutrons, having no electric charge, have an easier time getting close to a target nucleus like uranium than most other candidates. Most particles massive enough for this job carry some positive charge, and are repelled by any positively charged nucleus. So this was tried.

The hoped-for result was that uranium might be kicked into letting loose an alpha particle, which was the "customary" way in which borderline unstable nuclei made themselves smaller, more compact, more tightly packed, more stable. A kind of "induced radioactivity" was considered possible. Often in emitting an alpha or beta particle, a nucleus also sends along a gamma particle, a burst of electromagnetic energy. (See more on this in Part IV)

Uranium was targeted for bombardment. In 1939 Otto Hahn and Fritz Strassman bombarded uranium with neutrons.

What they thought might happen, didn't. But looking carefully at the uranium sample after it had been bombarded revealed that in the debris were particles considerably larger than any radioactive particles that had ever been seen. In fact, among the reaction products, they isolated some nuclei that were about half the size of the uranium nucleus.

Here were not little chunks that were chipped or knocked out of the uranium nucleus. It remained for Lise Meitner and Otto Frisch to interpret what Hahn and Stassman had seen. The uranium had relieved itself in a much more massive way than by emitting a radioactive particle. It had split.

What happened came to be understood as a breaking apart, or "fission," of the entire uranium nucleus, with about half the nucleons going to one fragment and half to the other. The fragments are now known to be of great variety. There are several dozen ways in which the nucleons can divide themselves into pairs, but the split nucleus with the neutrons given off always add up to all the particles, though not all the mass, of the original uranium nucleus.

It was later found that most of the uranium in a natural sample does not fission when hit with a neutron. 99.3% of naturally occurring uranium is the ${}_{92}U^{238}$ isotope, which is radioactive, but does not fission. It is the 0.7% of the uranium that is the isotope that undergoes fission, ${}_{92}U^{235}$.

The fission reaction

Although there are many different possible pairs of fission products, the details of the differences among them are not important. We use one of the more common ones to represent "the fission reactions of ${}_{92}U^{235}$."

To find out how much energy might be available from such fission, we follow the example from Chapter 4, and begin by writing the reaction equation.

On the left hand side of the reaction equation, we have, first of all, a fissionable isotope, ${}_{92}U^{235}$. The uranium nucleus does nothing all by itself. Fission is activated by hitting the uranium nucleus with a neutron of fairly low energy. The neutron's energy can be said to lift the uranium nucleus over the energy barrier, that rim around the depression in the top of the mound.

Once the uranium nucleus is split, the fragments quickly clamp down because of the increased nuclear strong force of attraction among the nucleons in the fragments where distances are smaller. This clamping down tightens the bonds, decreasing the average bond energy remaining, as the graph of Fig 3.2 shows. To help stabilize the new fragments, three neutrons are discarded, rather vehemently, carrying with them much of the kinetic energy released with the tightening of the bonds.

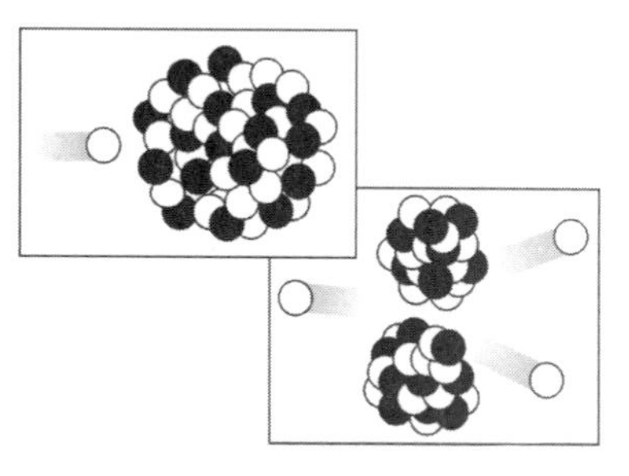

Fig 5.1 Uranium-235 splits into two smaller nuclei, in which the "packing" of the nucleons is tighter, releasing energy.

Here is the reaction:

$$ {}_{92}U^{235} + {}_{0}n^{1} \rightarrow {}_{56}Ba^{139} + {}_{36}Kr^{94} + 3\,{}_{0}n^{1} + \text{Energy} \qquad [5.01] $$

The fragments in this typical fission reaction are nuclei of Barium and Krypton. The fragments are not the naturally occurring isotopes of Barium and Krypton, but are unstable isotopes themselves, and over time, go through a series of radioactive dissociations that contribute about 6% of the eventual total energy released by uranium fission.

The energy of the fission itself is carried away as kinetic energy by the three very high speed neutrons. In a reactor, these neutrons can be robbed of their excess energy by passing them through a material like water, that becomes heated as it slows down the neutrons. The heated water can then be used to generate steam to drive a turbine that is connected to an electric power generator, which sends the energy to the user in the home or factory. In this part, we will follow the energy through the various steps to the end user.

The kinetic energy that the neutrons carry away has a certain amount of mass, and consequently the total mass of the fragments and the three neutrons is less than the mass of the original uranium nucleus plus the activation neutron. By the same methods of calculation that were used in Chapter 4, it is now a routine matter to use the mass defect in the reaction equation as a measure of the energy yield of this reaction.

The energy yield of uranium fission

Substituting the isotope masses of the constituents in [5.01], we obtain the following equation per mole of fissioned uranium,

$$\begin{aligned} 235.0439\,\mathrm{g}/N + 1.00867\,\mathrm{g}/N \\ = 138.905\,\mathrm{g}/N + 93.915\,\mathrm{g}/N + 3\times(1.00867\,\mathrm{g}/N) \\ + \text{Energy} \end{aligned} \qquad [5.02]$$

totalling the molar masses on each side, gives,

$$236.05257\,\mathrm{g}/N = 235.84601\,\mathrm{g}/N + \text{Energy} \qquad [5.02a]$$

$$\text{Energy Yield} = 0.206563\,\mathrm{g}/N$$

This can be expressed in energy units,

$$\text{Energy Yield} = (0.206563\,\mathrm{g}/N)\times(1\mathrm{kg}/1000\mathrm{g})\times(3\times10^{8}\mathrm{m/s})^{2}$$

$$= 1.859 \times 10^{13}\text{Joules/mole of uranium}$$

For practical use, this is better expressed "per gram" rather than "per mole."

$$\text{Energy yield} = (1.859 \times 10^{13}\text{J/mole}) \times (1\text{mole}/235\text{g uranium})$$

$$= 7.91 \times 10^{10}\text{Joules/gram of uranium}$$

$$= 21{,}970 \text{ KWHr / gram of uranium} \qquad [5.02b]$$

About Watts and Kilowatt Hours

1 Watt = 1 Joule/s (defined) is a unit of Power; rate of doing Work

1 Joule = 1 Watt·sec (Energy)

1000 Joules = (1000 Watts)·(1 sec) [used by a 1000W bulb in 1 sec]
3,600,000 Joules = (1000 Watts)·(3600 sec) = 1 KW Hr (Energy)

approximate daily home usage of electric energy: 20 KW Hr

(1 calorie = 4.18 Joules)

A comparison can now be made between the energy yield of uranium fission and combustion of gasoline:

1 gal gasoline yields about 42 KW Hr ≈ *2 days'* home energy demand

1 g of uranium yields 22,000 KW Hr ≈ *3 years'* home energy demand.

The reaction in [5.01] (with numerous variations that produce other fission fragments) is the energy source in working reactors throughout the world today. Almost all reactors in the United States use uranium as a fuel.

The focus in this book will be on the pressurized water uranium reactor that is the most common technology in U.S. reactors. In Part III we will deal briefly with other technologies and the plutonium breeder reactor.

We will examine next the activation process in the reaction in which uranium is split (fission).

Problems

5.1 List and describe the activation of the following processes.

(1) burning candle
(2) emptying a water bucket
(3) Parachuting from an airplane
(4) Falling array of dominoes set up so that each domino knocks over its neighbor
(5) "infectious laughter"
(6) spreading a rumor
(7) growth of a tree in a forest
(8) popping corn

6.
How to keep the log burning

A wooden log does not burn until it is lit. That is because the flammable part of the wood does not combine with oxygen at room temperature. Lighting the log is an activation step. We are fortunate that this step is required, because otherwise there would be no wood for us to burn; it would all have burned long ago.

Activation can be costly. If the log went out every time we stopped lighting it, we would soon run out of matches and kindling, and in the meantime, we might be investing more energy in keeping the log lit than we get from its burning. The same thing is true for the activation of uranium fission by collision with a neutron.

Fortunately, once the log is lit, the fire of the log will continue to light new parts of the log, until it is all burnt up. Some of the energy output from the burning log is used to activate the burning of new parts of the log. This is referred to as a chain reaction.

The same is true in uranium fission. Like matches, a few neutrons are easy and cheap to come by (actually there are always a few stray neutrons around that cost nothing). But enough matches to keep lighting the new parts of the log, and enough neutrons to keep activating the fission of all the uranium nuclei, are both costly and cumbersome.

Fortunately, uranium fission can be a chain reaction. The output of the fission of a uranium nucleus includes the emission of three neutrons [see eq 5.01]. If even just one of the three is able to activate fission in one other uranium nucleus, the process will continue until all the uranium has split. The fissions that are activated by the neutrons emitted in a previous fission are described as being of the "next generation" of fissions. The ratio of next generation fissions to original fissions is called the *generation ratio.*

If the *generation ratio* is at least one, each fission will activate at least one other one, and uranium fission will be a chain reaction. Once it has been started, it will continue. Because there are more neutrons produced than are needed in further activation, the process can quickly get out of hand. Unless this is desired, as it is in the case of a uranium bomb, this creates a stability problem.

This raises the question, "Why doesn't it get out of hand? A few stray neutrons are always around from a variety of sources, so it would seem that not only is uranium fission a self-starting reaction, but every little piece of uranium should blow itself up without help from outside." Recall that the fission energy of one gram of uranium is equivalent to the chemical energy of 525 gallons of gasoline.

There are really three questions. The first, why doesn't it get out of hand? the second, since apparently it doesn't even go, how can we make it go? and the third, if we can make it go, can we stop it?

Critical mass

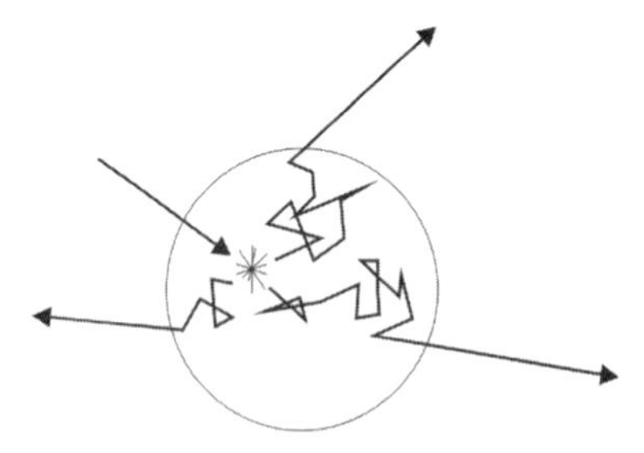

Fig 6.1 All three neutrons get away.

The answer to the first question is that in a small sample, fewer than one out of the three neutrons will find another uranium nucleus before exiting the sample. This is illustrated in Fig 6.1, where all three neutrons produced in a fission leave the sample without activating another fission. Occasionally one neutron will produce another activation, but that is not enough to sustain the process.

The remedy for this situation is almost too obvious. As the three neutrons produced in each fission bounce around and eventually leave the sample, the likelihood that any one of them will make a successful hit on a new nucleus increases with the distance that it travels through the uranium sample. The longer it stays in the sample, the more chances there are that it will make a successful hit on another uranium nucleus.

The way to make the chain reaction go, then, is to make the chunk large enough [Fig 6.2] so that the odds favor at least one new fission from each set of three neutrons emitted, making the generation ratio equal to or greater than one. The "critical mass" of uranium is that amount which is large enough to sustain a chain reaction.

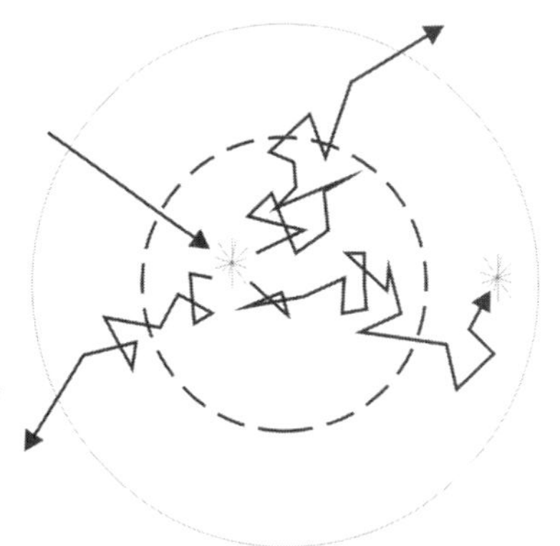

Fig 6.2 If the chunk is large enough, one or more neutrons will cause a subsequent fission.

How much uranium makes the mass "critical?" For the early bomb scientists this was a hard question. Such a critical mass would indeed be a self starter, and would explode as a bomb. So the critical mass can not very well be determined by trial and error. Experiments in a "pile" consisting of uranium mixed with graphite and rods of other materials that gave operators some measure of control allowed experimenters to produce for the first time a self-sustaining reaction, without actually blowing up the nearest city.

Still, partly because a number of other factors, such as shape, density, and the presence of neutron-absorbing material are also involved, determination of what constitutes critical mass was one of the key technical problems in making the first bombs. It is still a matter of dispute among historians whether it was a purposeful error made by scientists in the German bomb project not eager to give the Nazi state that weapon, or whether it was a genuine mistake, but it is believed that the German project was abandoned because of a belief that the critical mass is so great as to make building a bomb impractical.

If it is shaped properly, critical mass is approximately 5 kg of fissionable uranium, the isotope $_{92}U^{235}$. This amounts to a chunk that weighs about 10 pounds, and could be held in one hand were it not, because of the chain reaction, dangerous to do so.

The Stability Problem

There is no stability problem in building a nuclear bomb. It is built out of sub-critical (less than critical) components. If the mass is less than critical, the generation ratio is less than one. If the mass is even slightly greater than critical, the generation ratio is greater than one.

Because the reaction dies out very rapidly when the generation ratio is less than one, a chunk of uranium that is close to but not of critical mass, is not hazardous, and will not sustain a chain reaction.

For reasons that we will explore in Chap 8, an amount of uranium that is even slightly above critical mass will fission almost all of its uranium in a few thousandths of a second.

You will see how, in principle, the design of a bomb is a fairly simple matter once the question of critical mass is solved. But, when we try to generate electricity from nuclear energy, we need to sustain a chain reaction that does not explode in a bomb. Instead, we want it to maintain fission at a steady rate, minute after minute, hour after hour, day after day. This is much more difficult than building a bomb.

Problems

6.1 For the processes of Problem **5.1**, identify those which are chain reactions, and, for those that are, describe their "criticality."

7. Designing a Bomb

Once accurate data have been obtained for determining critical mass, two major technological design problems remain.

The first is "enrichment" (partial purification) of the fissionable portion of the uranium ore that comes out of the ground in uranium mines. The other is speed. The two are related.

Enrichment

Only 0.7% of the naturally obtained uranium is the fissionable isotope, ${}_{92}U^{235}$. Someone accustomed to purification methods in chemistry will be surprised at the great difficulty of separating ${}_{92}U^{235}$ from ${}_{92}U^{238}$.

Generally chemical purification relies on the difference between the chemical properties of the desired material and those of the impurities. These include melting and boiling temperatures, reactivity with acids and bases, crystallization, etc. Water, for instance, can be separated from its mineral content by distillation; at 100° the water boils and leaves the rest behind. Copper sulfate can be separated from impurities in solution by growing a gorgeous blue crystal, because most of the impurity ions do not fit into the crystal structure of copper sulfate.

This is not to belittle the skill needed to perform chemical separation, but to appreciate that isotope separation is a great deal more difficult.

All the chemical properties of two isotopes of the same element are the same. The two isotopes of uranium melt and boil together, they dissolve in and crystallize from solutions together, they react with acids, oxygen, and other reagents exactly the same. This is because all these properties are determined by the electrons that surround the nucleus, and not by the nucleus.

The only difference between the two uranium isotopes is in the mass of the nucleus. The masses of the atoms differ by a little over one per cent. But that is all.

Fortunately complete purification of the fissionable isotope, ${}_{92}U^{235}$, is not required. It is not necessary, either in bombs or reactors, to use uranium that is pure ${}_{92}U^{235}$. It is sufficient that it be "richer" in the fissionable isotope than the meager 0.7% that it is in the naturally occurring mineral. And so one speaks not of purification, but of "enrichment."

Enrichment is accomplished in stages which sort the isotopes by marginal preferences that they show in one of three categories of selection processes. In each stage, there is some enrichment, and by sufficient repetition of the stages, varying degrees of enrichment are possible.

The first method used depends on the tendency of heavier particles to sediment faster than the lighter particles in a centrifuge. This is like throwing a bit of beach sand into a glass of water. The large grains settle to the bottom first, leaving the fine grains suspended in the upper part of the water for a longer time. But, if the grains of sand were by only 1% different in size, the differentiation of the larger and the smaller by this method would require repeated sedimentation. Centrifugation of uranium salts is similar to the sedimentation of grains of sand.

The second method used involves producing a relatively low boiling compound of uranium, vaporizing it, and diffusing the gas through a porous filter. The lower mass isotope tends to diffuse slightly more rapidly than the heavier one.

Ionization by a tuned-laser method is the most recent method developed.

None of the three methods is cheap, quick, or efficient. The technology of enrichment requires costly, elaborate and, in practice, physically large, installations, and has been an impediment (fortunately) to the spread of nuclear weapons technology. Reactors can use uranium that is enriched with $_{92}U^{235}$ to only about 3%, while weapons require uranium that is about 90% fissionable. It is therefore possible to sell, for electric power generation, reactor grade uranium, which is not suitable for making bombs.

Speed

The basic ingredient of a bomb consists of two pieces of uranium, a main part that is about 90% critical, and a second portion that provides enough additional fissionable uranium so that, together, the two parts are above critical mass. By keeping the two parts separate, the bomb can be built, transported, and dropped without exploding. At the chosen moment, by the use of a timer, or an altitude sensor, or a proximity fuse, or by remote control, the two parts are joined together, and the bomb explodes.

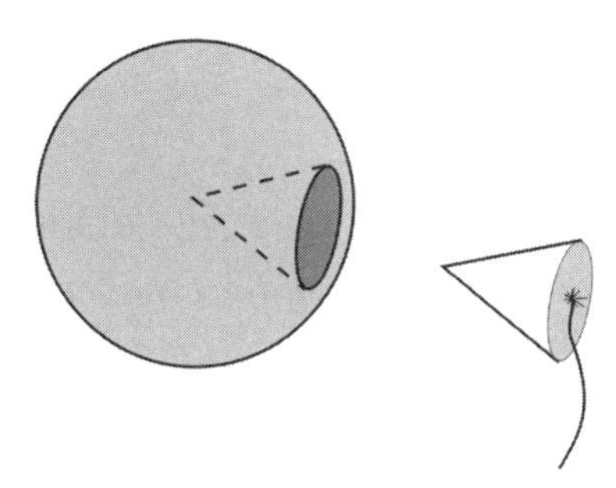

Fig 7.1 Concept drawing of a uranium bomb. Cone is fired by an explosive lit by a fuse. Together with main portion, critical mass is reached.

If the joining of the parts is not sufficiently rapid, or if the geometry of the resulting critical mass does not permit the fission to proceed rapidly enough, the bomb will explode "gradually," causing most of the uranium to scatter without having fissioned, dispersing the uranium in small chunks which are sub-critical (less than critical) in size, and whose nuclear energy remains locked up in the intact uranium nuclei that simply fall to the ground.

The phrases "rapidly enough" and "gradually" are used in reference to an incredibly tight schedule. If most of the potential fission has not occurred by the first micro-second, the bomb will explode before all the uranium has given up its explosive energy.

It takes time for the second fragment to make intimate contact with the main part of the bomb, and so this time must be minimized. In Fig 7.1 we have drawn, schematically, the second part in the form of a cone that is fired by a conventional explosive into a slot in the main bomb piece.

Once fission begins, the chain reaction proceeds at a rate that is limited by the time it takes the neutrons emitted from one fission to make successful collisions with other fissionable nuclei to sustain the reaction.

For this to occur fast enough, the physical dimension of the bomb must be as compact as possible, and this requires that as little space as possible be occupied by non-fissionable uranium. This is why weapons-grade uranium is enriched to contain about 90% fissionable uranium.

The proliferation of fission throughout the bomb is enhanced by the use of an "initiator," a neutron source that is "turned on" at the moment when the two parts of the uranium mass are joined. The initiator gets things going fast by supplying as much as 10^{20} neutrons.

The average amount of time that elapses in a bomb between one fission and the subsequent fission that is produced by one of the emitted neutrons in the original fission is 1×10^{-8}sec, or about 0.01 microsecond. This time is called the "generation time." This means that there is time for about 100 generations. In the next chapter we will develop the idea of generations and their relation to explosion speed in bombs and to stability in reactors.

8.
Generations: mothers and daughters

We have already gotten a glimpse of the tight time scales that describe what happens when a chain reaction proliferates fission throughout a mass of uranium that is super-critical (greater than critical). In a bomb, we found it important that fission take place so rapidly that all the uranium in the bomb will undergo fission before the explosion scatters that material.

In a reactor, we have different goals. We want to arrange things so that what is desired in a bomb, does *not* happen. We want most of the uranium to still be there in one micro-second; in fact we want most of it to still be there tomorrow. Yet we do want fission to take place, because it is the purpose of the reactor to draw from uranium fission the energy that produces the electricity that is sent from the reactor plant to homes and factories.

It turns out that the task of making everything go fast enough is like child's play compared to the task of making it go just at the right rate so that it will keep producing energy at a steady rate for months and years, and *not* blow up.

We must explore the means that might be available to control the approach to criticality.

The only way that we have mentioned so far is to add little pieces of uranium to a chunk of uranium, so that, by making that chunk bigger we gradually increase the likelihood that the emitted neutrons will have successful collisions that split other uranium nuclei.

We understand what "critical" means. It means that of the three neutrons that are emitted from the uranium nucleus during fission, at least one must produce another fission. But then, what describes "almost critical?" How do we define how close to critical we are? Other than adding little pieces onto a chunk of uranium until it blows up, what devices are there for reaching a condition that is so close to critical that it maintains a steady state in which fission continues, without reaching critical mass and causing an explosion?

We will find that, regardless of the tricks we can discover (and there will be some) for manipulating the approach to criticality, it will be a daunting assignment to keep a critical mass stable. We will find out that once we pass across that boundary past critical mass, the bomb condition looms almost immediately.

We now develop two concepts that will enable us to describe what so far appears rather vague. Both concepts deal with the chain reaction as a succession of generations.

Generation time

We know what a generation is in the life of humans. It is the coming of age of children until they reach the stage in their lives when they play the role of their parents. The *generation time* is the time elapsed between the birth of children and the time that they have children themselves.

Between parents and their children is one generation. From grandparents to parents is one generation; from grandparents to children is two generations. You are four generations removed from your great-great-grandparents. A generation of humans is often estimated at 25 years.

A generation of cats can be less than a year. A generation of amoeba is a few hours.

What do we mean when we say, for example, that the generation time of nuclear fission in a uranium reactor is 2 μsec? (The Greek letter μ, pronounced "mew," means "one millionth"; so 2 μsec means 2 millionths of a second, or 2×10^{-6} sec.)

Suppose that at a time, t = 0, we expose the sample of uranium to an initiator that causes 10,000 nuclei to undergo fission. In those initial fissions, 3 neutrons are emitted from each nucleus, for a total of 30,000 neutrons. Suddenly 30,000 neutrons go dashing about looking to hit another uranium nucleus to split. If this sample of uranium is exactly critical, 20,000 of these 30,000 neutrons will get away, and 10,000 of them will find a new uranium nucleus and cause it to split, within 2 μsec. Now, there are 10,000 new fissions, 30,000 new neutrons looking for uranium nuclei to split, 20,000 get lost, and again, within 2 μsec, 10,000 of these neutrons cause 10,000 new fissions, and so on. A chain reaction would be just exactly maintained. Each 2 μsec, a new generation of fissions takes place.[1] If, as we have described it, an emitted neutron takes an average of 2 μsec to produce another fission, the *generation time* is said to be 2 μsec.

Generation ratio

In each generation of humans, two adults have an average of two children. In each generation of cats, there may be 6 for each mother cat.

In the fission of uranium, the 3 neutrons emitted from each fission may give rise in the next generation to the fission of anywhere between 0 and 3 further nuclei. If all three neutrons find

[1] It needs hardly be said that this picture of fissions taking place in lock-step is not the way it really happens. It is more nearly a continuous process. This does not affect the validity of the analysis presented here.

uranium nuclei to fission, the generation ratio is 3. If one out of 3 finds a uranium nucleus to fission, the generation ratio is 1.

Because we are not discussing a single uranium nucleus and its three daughter neutrons, but large collections of uranium nuclei, these ratios need not be 0, 1, 2, or 3, but may be fractional quantities that represent averages.

Suppose that in another example, out of the 30,000 neutrons emitted in 10,000 nuclear fissions, 21,000 get lost, and only 9,000 new fissions take place. The next generation has only 0.9 as many fissions as the first. A multitude of conditions in this example determined that the likelihood of a neutron producing a new fission will be 9,000/30,000, less than 1 out of 3. The generation ratio is (9,000 fissions)/(10,000 fissions), or 0.9. There will not be a sustained chain reaction.

What will happen in that case? There is a second generation of fissions, and a third, and many more. But each generation is smaller than the one before, by the ratio of 0.9. And so the first generation is $(10{,}000)\times(0.9)$, or 9,000. (We call the original generation the "zeroth" generation, for reasons that will soon become apparent.) Now the second generation is produced from a smaller first generation, and is again smaller by the ratio of 0.9. The second generation is $(9{,}000)\times(0.9)$, or 8,100. This could also have been written as $(10{,}000)\times(0.9)\times(0.9)$ or $(10{,}000)\times(0.9)^2$.

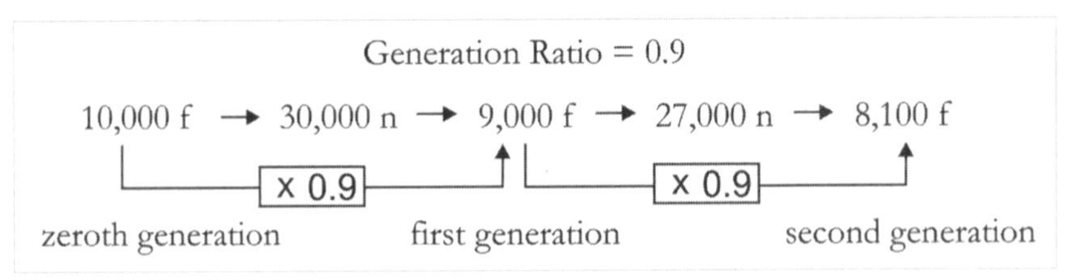

Fig 8.1

Using the notation that (0.9) multiplied by itself 15 times is written $(0.9)^{15}$, we can see that the 15th generation consists of

$(10{,}000) \times (0.9)^{15}$ fissions, or $(10{,}000) \times (0.2059)$ which is 2,059 fissions. You see that the chain reaction has not died out yet, even after 15 generations. But it is on its way.

How to do this on your calculator
To calculate $(0.9)^{15}$ on your calculator use the [x^y] button
1. Punch in [0.9] 2. Punch [x^y] 3. Punch [15]

You can verify that (in the example above, where we start with 10,000 fissions and the generation ratio is 0.9) the number of fissions in the 50th generation is 51, and in the 100th generation, it is less than 1. Now it may seem that waiting 100 generations for this chain reaction to die down is a long time to wait; recall that one generation may be as short as 0.01 μsec, so that 100 generations takes just one millionth of a second.

The general equation for the number of fissions, f_n, in the nth generation is

$$f_n = (f_0) \times (r_G)^n \qquad [8.01]$$

Notation used:	fissions in zeroth generation	f_0
	fissions in nth generation	f_n
	Generation Ratio	r_G

There is, finally, the obvious relation between the generation time, t_G, the time elapsed, t, and the number of generations, n. If one human generation is 25 years, then the time of four generations is (25yr) × (4), or 100 years.

In general, then, in nuclear fission generations as in human generations,

$$t = (t_G) \times (n) \quad [8.02]$$

which is more often used inverted, that is, used for answering the question, *how many* generations occur in a time interval, t, if the generation time is given? The inverted relation is,

$$n = (t) / (t_G) \quad [8.03]$$

Example

How many generations occur in $10\,\mu sec$ if the generation time is $0.01\,\mu sec$?

$$n = (t) / (t_G) \quad [8.03]$$

$$n = 10 \times 10^{-6} sec \; / \; 0.01 \times 10^{-6}\ sec$$

$$n = 1000$$

Problems

8.1 If you start with a zeroth generation of 10,000 fissions, and the generation ratio is 0.8, how many fissions will there be in the first generation; how many in the second?

8.2 If you start with a zeroth generation of 10,000 fissions, and the generation ratio is 0.8, how many fissions will there be in the 15th generation; how many in the 16th?

8.3 If the generation time is $40\,\mu sec$, how much time has elapsed from the start to the 15th generation?

Ans: **8.1** 8,000; 6,400 **8.2** 352; 281 **8.3** $600\,\mu sec$, or 0.6 msec

9. The squeeze between boom and bust

You observed in the example used in chapter 8 that, with a generation ratio of 0.9, the number of fissions went from 10,000 in the zeroth generation, to 9,000 in the first generation, to 2,059 in the 15th generation, to 51 in the 50th generation, and then to less than 1 in the 100th generation. These numbers came from equation [8.01],

$$f_n = (f_0) \times (r_G)^n \qquad [8.01]$$

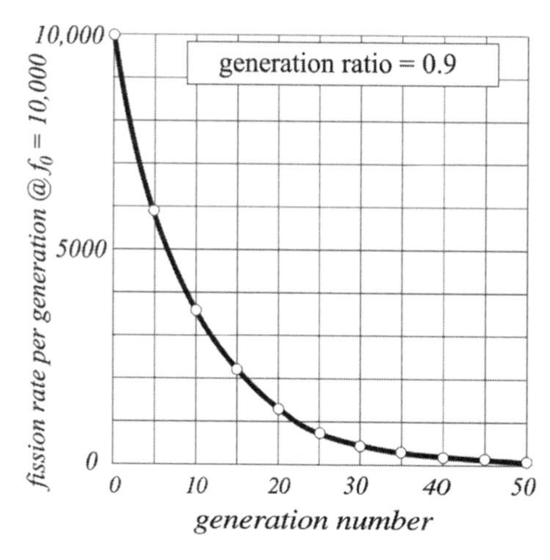

Fig 9.1 Fission rate per generation

In this equation, "n" is the number of generations. In this example, f_0 is 10,000, r_G is 0.9. f_n would be f_{15} in the 15th generation and its value would be 2,059. A graph of this function is shown in Fig 9.1.

1. Generation ratio is less than one

It turns out that the *shape* of this function is generic. It is called an exponential decay function, and applies to these fission problems whenever the generation ratio is less than one. Physically, generation ratio of less than one means there are fewer fissions in every succeeding generation, by the ratio, r_G.

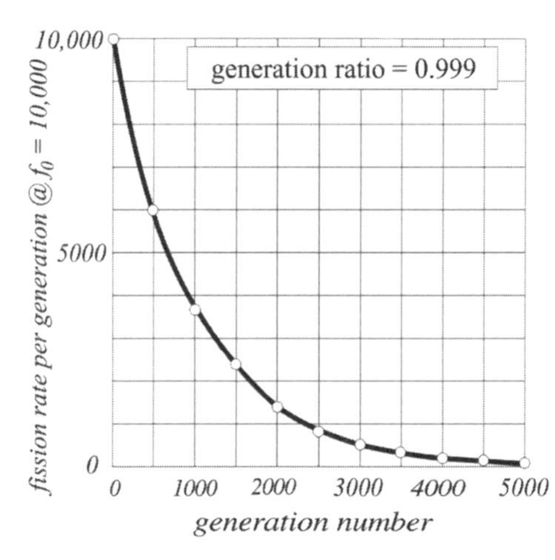

Fig 9.2 Fission rate per generation

If the generation ratio is ever so close to one, let us say it is 0.999, then the fission rate in successive generations diminishes much more slowly, and one might say the result is almost what we want, a process that keeps on going day after day. But the shape of the decline in fissions remains the same. Fig 9.2 shows a graph of the function of Eq 8.01 with a generation ratio of 0.999.

The generation time in reactors is typically of the order of μsecs (millionths of a second) In Fig 9.3, the generation ratio is still 0.999, but the number of fissions per generation is expressed as a function of *time* instead of generation number. In Fig 9.3, the generation time is 20 μsec.

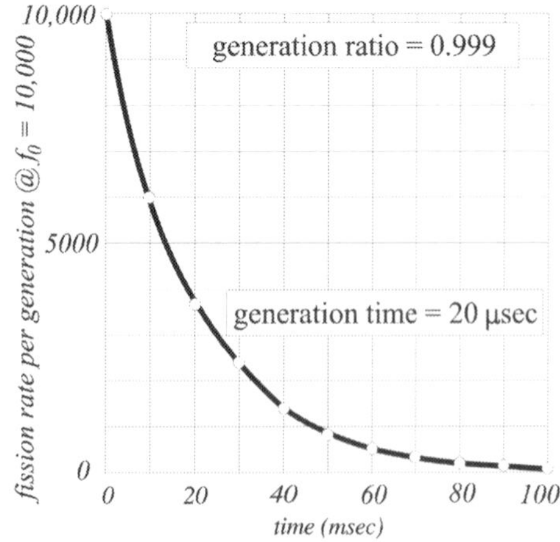

Fig 9.3 Fission rate per generation

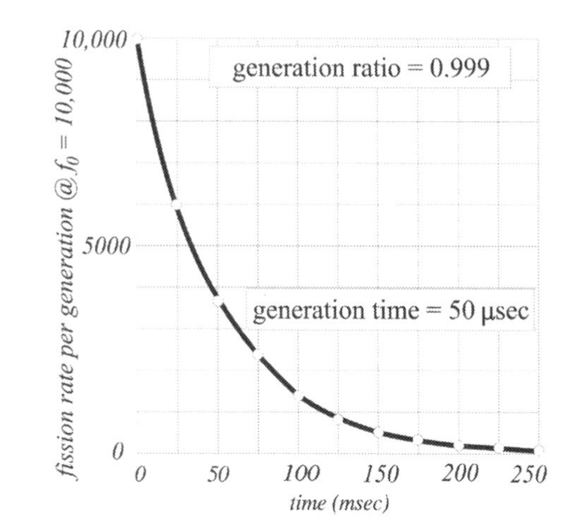

Fig 9.4 Fission rate per generation

If the generation time is longer, the shape of the curve remains still the same, except that the time scale is stretched. Fig 9.4 shows the decay function when the generation time is 50 μsec.

The fact is, that fission may stop in more or fewer generations, in a longer or shorter time, but almost always in a matter of much less than a second. It does not keep going day after day, not even second after second. If r_G is less than one, the time it takes for a uranium fission reaction to die out is never very long.

This mathematics, especially the persistence of the pattern shown by the shape of all the graphs of Figs 9.1-4, teaches us that the need to be able to achieve a chain reaction that does not die out can not be satisfied by inching the generation ratio closer and closer to 1.0000.

What then is the way to achieve a chain reaction that does not die out, but that also does not hand us a bomb?

Let us review the problem of stability. In order to keep a reactor running, we must have a chain reaction that sustains fission at a constant rate, to produce electricity day in and day out at a moderate rate (meaning no burst of energy, as in a bomb). To sustain the chain reaction, we know that we require that the fissions at least replace themselves from generation to generation. We have seen that even if they replace themselves at a rate of 0.999 (or 99.9%), the fission rate dies out within 10,000 generations, which can be much less than a second. The rate has gone from

10,000 in the zeroth generation to less than 1 by the 9,000th generation.

2. Generation ratio is greater than one

We know that if there is at least one new fission from the three neutrons emitted in an original fission, the chain reaction will continue. We will find, however, that in this case, the exponential function gives us a disaster in the opposite direction from the quick death that we found when the generation ratio is less than one.

We know what we need. We need a generation ratio of exactly 1.0000000. In that event, for each fission, exactly one out of the three neutrons emitted will produce another fission in the next generation. It means the rate of fission will continue steadily, at a constant rate, day after day, just as we would like it.

In the next chapter we will examine the technology which allows us to control the generation ratio by raising or lowering control rods in the reactor vessel. We will find that there is a way to adjust the generation ratio, and presumably that would solve the problem. In other words, we can sit at a console in the control room and turn a knob to set the generation ratio wherever we want it. Where we want it is at 1.000000.

But we will find that this is like using an automobile accelerator to try to adjust a car's speed to exactly 50.0000 miles per hour; it is not possible to control it that closely. We do well if we can keep the car going between 48 and 52 miles per hour. The adjustment that we have available to us in the reactor can do slightly better, but it will not allow us to control the generation ratio to be as close to 1.00000 as we would need. We ask, now, how closely do we need to keep the generation to 1.00000?

The equations 8.01 and 8.02 apply just as well when generation ratio is greater than one as they do when it is less than one. Their solution when the generation ratio is greater than one gives an exponential function that looks different from Figs 9.1-4.

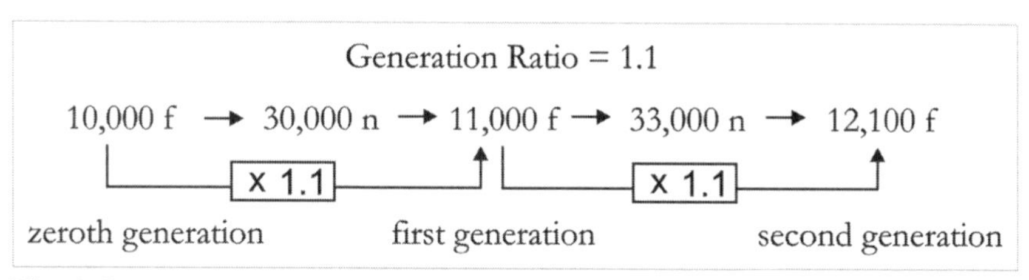

Fig 9.5

Fig 9.5 shows the sequence of the first two generation after the zeroth generation. In contrast to the numbers of Fig 8.1, the number of fissions now increases with each generation. It is now not a question of how long does it take for the number of fissions to die down, but how fast, and how high do the numbers get? There is no longer any limit, except the number of nuclei available to undergo fission. It is finished when the nuclei are all gone.

Let us look at what the exponential function does to us if, despite our best efforts in setting the dial, we aren't able to stay exactly on 1.00000, but let the generation ratio go up to, let us say, 1.1.

In that case, if we started with 10,000 fissions, after one generation we would have 11,000 fissions, after 15 generations we would have (use eq [8.01]) $(10{,}000)\times(1.1)^{15}$, or 41,800 fissions.[2] After 50 generations, we would have $(10{,}000)\times(1.1)^{50}$, or 1,170,000 fissions, and after 100 generations, $(10{,}000)\times(1.1)^{100}$, or 138,000,000 fissions. With the generation time of 1×10^{-5} sec, 100 generations is still just 1/1000 of a second. In 1/100 second, there are 1000 generations, and the number of fissions is

[2] The use of Eq [8.01] gives us the number of fissions in the n-th generation; it is the "rate of fission" per generation at that time. For the purpose, it is important to know the cumulative total of fissions that have occurred up to that time. This would be given by a sum of terms, $f_1 + f_2 + \ldots + f_n$, also written as $\Sigma_{i=1 \text{ to } n}(f_i)$, which can be evaluated as an integral. The function that gives that result is given by $f_{CUM} = f_0\,\{[r_G^n - 1] / [r_G - 1]\}$. The difference is barely noticeable in the graph..

$(10{,}000) \times (1.1)^{1000}$, or 2.47×10^{45} fissions, which is enough to fission every nucleus in 160,000,000,000,000,000,000 tons of uranium, far more than there is on the earth.

In Figs 9.6-7 we have focused on how much uranium has undergone fission after a certain amount of time. 5 kg of ${}_{92}U^{235}$ contains 1.27×10^{25} atoms.

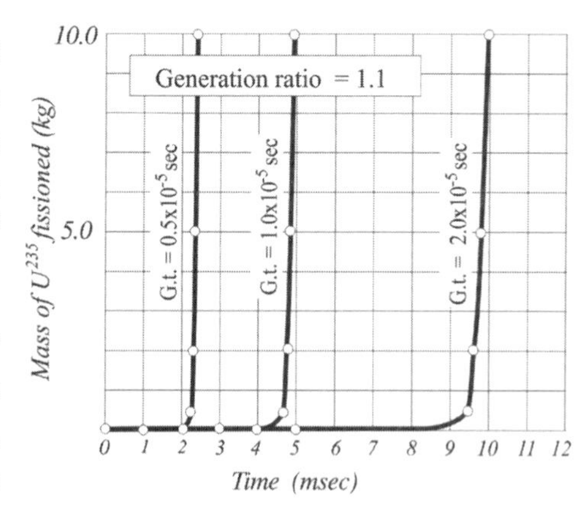

Fig 9.6 Mass of uranium fissioned

Fig 9.6 shows how fission progresses when the generation ratio is 1.1. Needless to say, we must be more careful than that. We can not afford to set the dial to where it gives us a generation ratio of 1.1. Let us be much more cautious, but remember that if we let the generation ratio go below 1, the chain reaction will die out very quickly. Let us turn that dial very carefully so that the generation ratio is just a wee bit above 1.000, let us say at 1.001.

Just as the shape of the function always remained the same when the generation ratio was less than one (Figs 9.1-4), the exponential functions now look the same with a generation ratio of 1.001 as they did when the generation ratio was 1.1, except shifted to a different time scale. Furthermore, changing the generation time also leaves the shape of the function unchanged, only shifted on the time axis.

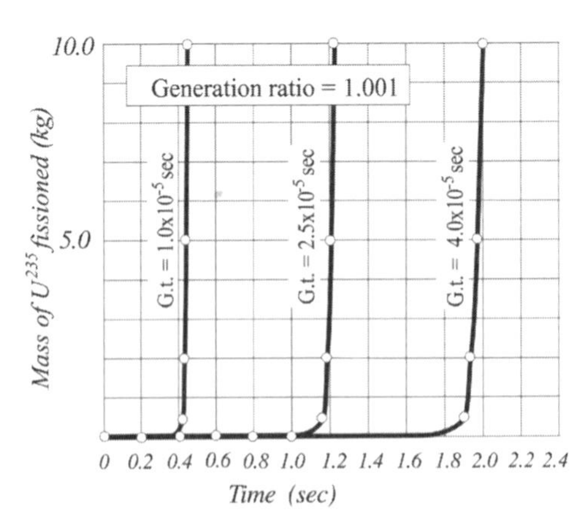

Fig 9.7 Mass of uranium fissioned

With f_0 = 10,000 fissions, generation ratio of 1.001, and a generation time of 1×10^{-5} sec, the rate of fissions after 1/1000 sec (100 generations) is still only 11,000. But after 1/100 sec, which is 1000 generations, the number of fissions is 27,000. After 1/10 second, which is 10,000 generations, there are 219,000,000 fissions, and after 1 second, the number of fissions has grown to 2.5×10^{47}, again far more than all the uranium on this planet.

The exponential function, when generation ratio is greater than one, can be described as a function that, on a scale of kilograms of uranium fissioned, does nothing for some period of time, and then suddenly shoots up and goes off the scale. The shape remains always the same; only the scale of time varies when the generation ratio changes, and when the generation time changes.

Being exceedingly careful has netted us 1 second of time, rather than 1/100 second, to get that dial down before we blow up all the uranium we have in our reactor. And setting a dial to 1.001 as opposed to 1.000 is probably not easy to do in the first place.

Recall that if we very carefully adjust the dial so that the generation ratio is 0.999, it would be just 1/10 sec before the chain reaction had died completely, with the number of fissions (from an original 10,000 per generation) down to below 1 per generation.

There is little prospect that the fission of nuclear fuel can be controlled by adjusting the control rods. It is humanly impossible to control the generation ratio to sustain a chain reaction that does not either die out or fission all the uranium in the reactor in less than a second.

One might conclude that the task of building a safe, working reactor is beyond technological possibility. It is our luck that nature comes to the rescue, and provides us with a built-in self-regulation system that makes it possible after all.

Problems

9.1 Fill in the missing information in the table below.

The conditions are:

Generation time (t_G) = 1.0×10^{-8} sec;

Zero generation fissions $f_o = 1$ (no initiator is used)

Time	n	f_n @ $r_G = 1.001$	f_n @ $r_G = 0.999$
0.01 msec (=1×10^{-5}sec)	1000	2.72	0.368
0.1 msec (=1×10^{-4}sec)	10,000	21,900	
0.2 msec (=2×10^{-4}sec)	20,000		---
0.3 msec (=3×10^{-4}sec)	30,000		---
0.4 msec (=4×10^{-4}sec)			---
0.5 msec (=5×10^{-4}sec)			---
0.6 msec (=6×10^{-4}sec)			---
0.7 msec (=7×10^{-4}sec)			---
0.8 msec (=8×10^{-4}sec)			---
0.9 msec (=9×10^{-4}sec)			---
1.0 msec (=1×10^{-3}sec)			---

Estimate graphically the time at which 5kg of U^{235} has been fissioned.
{5kg is about 20 moles of U^{235}, which is about $(20)(6\times10^{23})$, or 1.2×10^{25} fissions.}

Graph f_n as a function of time for r_G=1.001 from time=0 to time=1.0 msec.
Use a vertical scale for f_n going from 0 to 2×10^{25}.

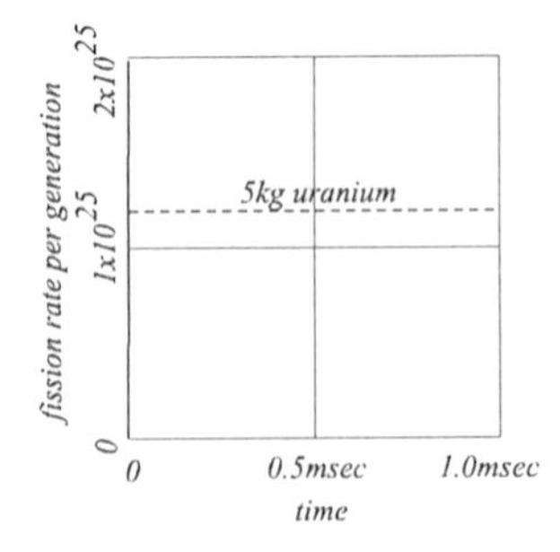

10.
Nature's nuclear thermostat

A thermostat is a device that is set to a certain desired temperature. Once set, it turns the heat on when the temperature is colder than the set point, and turns the heat off when it gets a little warmer than the set point.

The thermostat consists of a temperature sensor that reads the temperature and compares it with the set point temperature. Then, depending on whether it is lower or higher, it switches the heat on or off. This combination of features comprises a "self-regulating feedback mechanism." It does its job all by itself. It does not ring a bell when the switch needs to be turned one way or the other, to call you to attend to the switch. It does it all by itself.

Wouldn't it be wonderful if such a self-regulating feedback mechanism could be devised that would automatically crank up the generation ratio when the reactor is slowing down, and (very quickly) crank down the generation ratio when the reactor is starting to act like a bomb.

We saw in the last chapter that a reactor is extremely sensitive to very small deviations from the needed generation ratio, going off as if it were a bomb when the generation ratio is 1.001 and shutting down completely when it is 0.999. A reactor's response to these small deviations is extremely fast – too fast for human intervention.

The neutron's rule: "Easy Does It"

It is almost too good to believe, but such a feedback mechanism, not only completely automatic, but also extremely fast, is built right into the rules that determine how effective neutrons are in initiating new fission, which is what generation ratio is all about.

This feedback mechanism is based on the seemingly upside down relation between a neutron's speed and its ability to induce fission in another uranium nucleus.

One might think that to be effective in producing a new fission, the neutron would want to make a direct hit on a uranium nucleus with as much punch as possible. That is, after all, the experience we have grown up with: to smash a bottle with a small stone takes good aim and a strong arm.

The way a neutron causes fission in a uranium nucleus is different from the way a small stone breaks a bottle. Although the whole process of inducing fission takes a very short time, it consists of a sequence of two steps. The neutron does not simply crack open the uranium nucleus on impact. It has to become a part of it first. The neutron is absorbed into the uranium nucleus first, making a new isotope of uranium,

$$ {}_{92}U^{235} \quad + \quad {}_{0}n^{1} \quad \rightarrow \quad {}_{92}U^{236} \qquad [10.01] $$

Only then does the new, unstable isotope ${}_{92}U^{236}$ undergo fission. Within about a trillionth of a second it divides into two fragments and releases three neutrons. Step [10.01] can be regarded as an activation step which requires the investment of energy brought to it by the bombarding neutron.

The significance of this activation step is that it is happens with greater probability if the neutron is going slowly than if it is going fast. The neutron is not so much a tiny bullet impinging on a large nucleus, as it is a wave passing over it. It takes a moment's time for that wave to become incorporated in the nucleus. If it is too energetic (too fast) it passes right over it without merging.

Moderator

Neutrons as they are emitted in fission, are far too energetic to be effective in producing further fission. They don't have the leisurely quality to become absorbed in the new nucleus as in [10.01]. To make neutrons better at inducing new fission, a material is placed in the reactor for the purpose of slowing the neutrons. This material is called the "moderator."

It is a material like graphite or water that contains low mass nuclei, like carbon or hydrogen. These nuclei exchange kinetic energy with the neutrons through repeated elastic collisions, until both the moderator nuclei and the neutrons have, on the average, the same kinetic energy. The kinetic energy of the nuclei of moderator is proportional to the moderator's temperature. The neutrons may be described as having a temperature, based on their average kinetic energy. As they mix, the temperature of both the moderator and the nuclei equilibrate, in the same way that in mixing cold and warm water the cold water is heated and the warm water is cooled until the mixture reaches an equilibrium temperature.

Whereas cooling the hot, fast, neutrons may be looked upon as a "moderating" influence, as far as it affects their ability to induce fission, it does the very opposite. In cooling the neutrons, it makes them vastly more potent as initiators of new fission.

If the moderator is water, then by circulating it through the reactor it can serve also to carry away the energy that it takes from the fast neutrons, and move it out of the reactor core to where it is used to generate electricity.

In a very short time, the fast neutrons coming from uranium fission undergo millions of collisions with the moderator nuclei. Neutrons that have in this way equilibrated their own temperature with that of the moderator are called, "thermal neutrons."

The word "thermal" does not refer to a specific temperature. It refers to the temperature of the moderator, so thermal neutrons are not always at the same temperature. Moderator that has been allowed to heat up to a higher temperature makes thermal neutrons that are warmer, and faster.

In this relation lies the secret of the self-regulation of the generation ratio. The hotter the moderator, the faster the neutrons, the less effective they are in inducing new fission, the lower is their generation ratio. There is a negative slope to the graph of generation ratio as a function of moderator temperature (Fig 10.1).

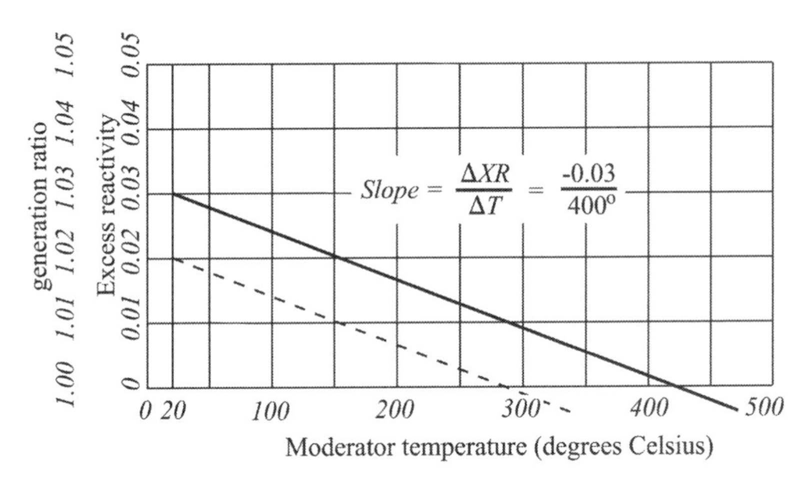

Fig 10.1 Graph of Excess Reactivity as a function of moderator temperature. An example.

The reader will notice that the vertical axis is labeled with two scales. The Excess Reactivity (XR)[1] is defined as the "excess of generation ratio above the needed 1.000." The translation is simple: $XR = r_G - 1$. The generation ratio needed to maintain stability in a reactor is 1.0000; the excess reactivity needed is 0.

---

[1] Yes, we know that the first letter of "Excess" is E, and not X.

In the example that is graphed here (solid line), the excess reactivity of that reactor at room temperature (20°C) is 0.03 (corresponding to a generation ratio of 1.03). The moderator is water. Water fills the reactor core, surrounding the fuel rods. If one could suddenly turn this reactor on at a (cold) water temperature of 20°, the fission would begin immediately to increase at a rate described by an exponential function (Fig 9.6) corresponding to a generation ratio of 1.03, and would be on its way to exploding within a fraction of a second.

Even faster than that exponential increase, however, is the heating of the water molecules. This is because the water temperature is raised by collisions of water with neutrons many times in each generation. As the water heats, the excess reactivity would quickly decline along the Excess Reactivity graph (solid line of Fig 10.1). As soon as the temperature reaches 420°, the excess reactivity would be reduced to 0 (generation ratio of 1.000).

At that point the self-regulation mechanism would hold the reactor at a nearly constant temperature, allowing small oscillations around 420°. If the temperature increased to 421°, the excess reactivity would dip below 0 (the generation ratio would decrease below 1.000) and the reactor would quickly shut down, the water would cool slightly, and its temperature might decrease to 419°. At that temperature, the excess reactivity would go slightly above 0 (generation ratio above 1.000) and the reactor would be again on its way to uncontrolled increase in fission, but for the fact that at 421°, the fission would again shut down, and the cycle would begin again.

The negative slope of the relation between water temperature and reactivity insures the self-regulation, and therefore the stability of the reactor.

The temperature at which the graph line of Fig 10.1 crosses the temperature axis is called the "operating temperature" of the reactor. By adjusting the control rods, it is possible to change the operating temperature. The "cold" excess reactivity in the example shown in the solid line of Fig 10.1 is 0.03. The dashed line representing cold excess reactivity of 0.02, is parallel to the solid line in Fig 10.1, but shifted downward, crossing the 20° temperature line at an excess reactivity of 0.02. The operating temperature in that condition is 287°, the temperature at which the dashed line crosses the temperature axis.

Operating Temperature

The slope of the line of Excess Reactivity as a function of temperature is called the Temperature Coefficient of Excess Reactivity (TCXR). Excess reactivity is more commonly expressed in per cent. In the example graphed in Fig 10.1, the slope is $-0.03/400^{O}$, or $-3\%/400^{O}$. Expressed in per cent per degree, that makes TCXR equal to –0.0075% per degree.

The equation,

$$TCXR = (\Delta XR)/(\Delta T) \qquad [10.02]$$

permits the calculation of operating temperature as a function of "cold" (20°) excess reactivity and TCXR. The condition is that at the operating temperature the excess reactivity is zero.

$$T_{OPER} = 20^{O} - XR_{COLD}/TCXR \qquad [10.03]$$

For example, if the TCXR is –.005%/degree, setting the cold excess reactivity to 1.5% would make the operating temperature 320°.

The method described in the box above shows how the fact of temperature dependence of generation ratio is useful not only because it stabilizes an otherwise very touchy nuclear reaction, but because it provides a way to adjust the operating temperature of the reactor.

The operating temperature determines how much steam is generated, which in turn determines how much electric power is generated.

Without this instantaneous and highly effective feedback mechanism, which was not designed by very clever humans, but was given to us as a gift of nature, it is hard to see how a nuclear reactor could have ever been built that has the stability to maintain fission at a constant and controlled rate day in and day out.

Problems

10.1 In a particular reactor, a chain reaction has been started by raising up the control rods to position 'A'. When the core temperature is 20°, and the rods are in this position, indicators in the control room show that for every 1000 neutrons in the reactor core, sufficient fission occurs to produce 1030 more neutrons.

(a) Express the 'EXCESS REACTIVITY' at 20° (in per cent).

(b) If the reactor reaches equilibrium at 470°, calculate the Temperature Coefficient of Excess Reactivity. To help you solve this problem, graph excess reactivity on the axes at the right.

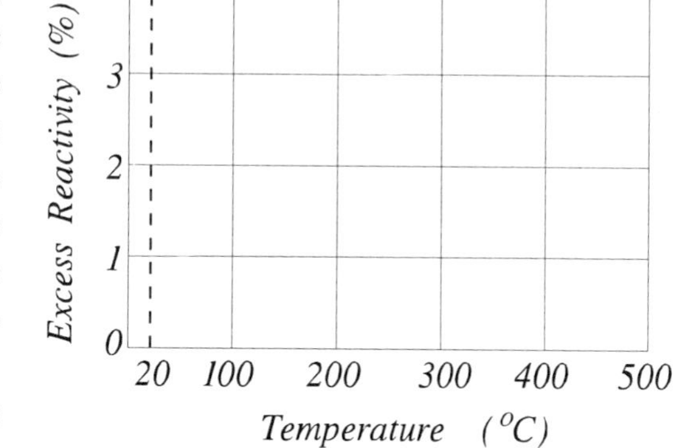

(c) It is desired to lower the operating temperature of this reactor to 320°. Draw a line in your graph parallel to the one from (b) but giving the desired operating temperature. Find the cold (20°) excess reactivity needed. Express it in percent.

Ans: (a) 3% (r_G=1.03) (b) –0.0067% / degree (c) 2%

11. Reactor Architecture I: The Core

It would seem that all the factors needed to build a reactor are now in place. These include (1) a nuclear reaction that yields lots of energy from every gram of fuel; (2) an activation process that keeps the reaction from happening without our say-so; (3) a means for triggering a chain reaction that can maintain its continued activation from its own by-product; (4) a feedback mechanism that maintains stability, keeping the chain reaction from either dying out or going out of control; and, finally, (5) a means for removing the energy produced, by way of heated water, to use by conventional means to generate electricity.

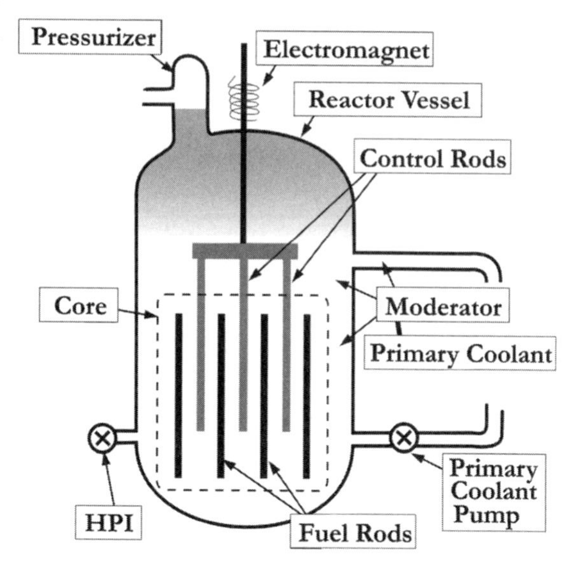

Fig 11.1 Schematic of the Reactor Core

We can now engineer these ingredients into a device that has been given the name, "nuclear reactor."

What we will describe here is a typical pressurized-water uranium fission reactor, that uses hot water as both moderator and primary heat transfer material.

The energy-producing portion of a reactor is in a region called the "reactor core." The core is inside a totally enclosed, steel-walled enclosure, called the reactor vessel. This vessel may be up to 50 feet tall, and is shaped somewhat like a huge thermos bottle, rounded on top and bottom for strength.

The core contains the uranium fuel, along with other equipment that is designed and controlled to keep an otherwise super-critical mass of fissionable material undergoing fission at a steady rate over a period of months or years between refueling.

The reactor vessel is filled wall-to-wall with water that is kept at high temperature and high pressure by the nuclear reaction. The volume of water is kept somewhat less than the volume of the reactor vessel, to allow a space at the top to be filled with steam, giving a measure of "softness" to the virtually incompressible water that might otherwise burst the 12-inch thick steel containment as it heats up. The steam is intended to be confined to a structure at the top of the vessel called the "pressurizer." The water level and the pressure in the pressurizer are continually monitored, and these data provide important information about the state of the reactor core.

All the fissionable material is in the reactor vessel. Once the reactor has begun operating, an increasingly large fraction of the original uranium has become nuclear waste in the form of highly radioactive fission fragments. It is these leftovers in the spent fuel rods that cause one of the biggest headaches for the nations that host nuclear power, because no one wants the stuff in their back yard. Thousands of tons of nuclear waste is in temporary storage, mostly submerged in water, waiting for a solution to the challenge of storing large quantities of isotopes, some of which will remain radioactive for thousands of years.

These fission products include the Barium-139 (${}_{56}Ba^{139}$) and Krypton-94 (${}_{36}Kr^{94}$) that appear in the reaction (Eq [5.01]) of chapter 5. These are not the common, stable and harmless isotopes of these elements, but are radioactive isotopes that dissociate over time, frequently into new isotopes which are themselves radioactive. This is true not only for barium and krypton, but also for most of the other fragments that result from uranium fission.

The mix of unused uranium and fission products is referred to as "spent" fuel. It remains confined in the reactor vessel, which in turn is inside a structure, usually cylindrical, shaped like a pill box some 100 to 150 feet tall, made of reinforced concrete 36 inches thick, called the "containment building."

The containment building looms large when it is viewed with a human figure standing at its base, but in aerial photographs or seen from a distance it is usually dwarfed by the giant inverted-tornado-shaped concrete structures called the "cooling towers." These cooling towers, we will discover, although huge and ominous looking, contain no dangerous structures or materials, and are among the safest places on a reactor site.

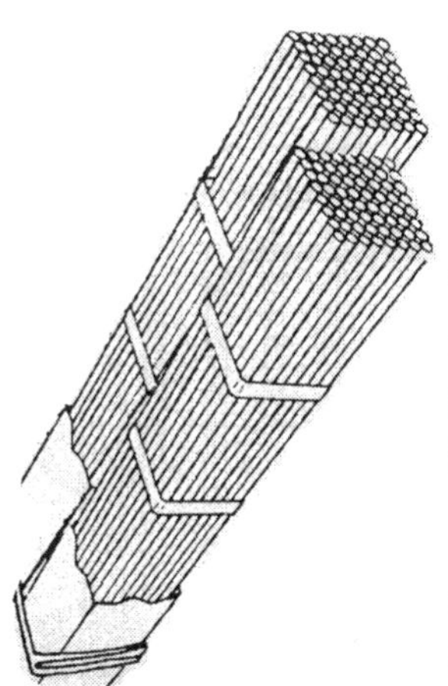

Fig 11.2 Fuel rod assemblies

The Core: 1. Fuel

For a variety of reasons, the uranium that is in the core as a fuel to undergo fission is not placed there, as might be imagined, in a large pile of chunks, like coal in a furnace.

The uranium is first placed in long steel-alloy tubes called "fuel rods," that are about 20 feet long and about an inch in diameter. The uranium typically is in the form of pellets of uranium oxide, in the shape of cylinders ½ inch in diameter and one inch high. These pellets are only mildly radioactive, and in sub-critical (less than critical) quantities and with reasonable precautions are not dangerous to be around or to handle.

The chemical composition is of no consequence. The only part of the uranium atom that is involved in fission is the nucleus; the electrons that form the bonds to the oxygen in the oxide of uranium play no role and do not interfere with the nuclear processes. The oxide is used primarily because it is a convenient form in which to place the uranium in the reactor.

Perhaps 200 pellets are dropped into each fuel rod. The fuel rods themselves are bundled and tied with steel bands into "fuel rod assemblies." One assembly may contain several hundred fuel rods. It is about one foot square in cross-section, and of the same length as the fuel rods. These assemblies are placed at some distance from each other in the core of the reactor.

Making the fuel "critical"

The uranium in the fuel rods is not highly enriched. Reactor fuel is natural uranium that has been enriched to about 3% from 0.7% fissionable ${}_{92}U^{235}$,. It is not necessary, nor desirable, to enrich it much beyond that level unless one needs a very rapid explosion, as in creating a bomb.

In order to achieve "criticality," the condition which sustains the chain reaction, the generation ratio has to be equal to, or greater than, one. Of every three neutrons emitted from fission, at least one must be able to initiate fission in another uranium nucleus. (It is essential, as we have already seen, that the generation ratio be equal to, but not greater than 1.0000, as in that case the exponential nature of the chain reaction would cause an explosion.)

What constitutes critical mass depends not only on how much ${}_{92}U^{235}$ is present, but on how it is distributed, and, on how successful we are in slowing neutrons down to the speeds at which they have a substantial chance of producing fission when they encounter another uranium nucleus.

We noted in Chapter 10 that a neutron encountering a uranium nucleus is more likely to be incorporated in it (which must happen first for fission to occur) if it lingers about the nucleus for a bit. If it is moving too fast, it is likely to swish right over the nucleus. It is this characteristic of the interaction between the neutron and the nucleus that gives us the negative temperature coefficient of reactivity, and allows us to operate the reactor without it teetering between blowing up and dying out.

If we do nothing to slow down the neutrons as they emerge from fission, they will be moving too fast, and will be most likely to escape from the core entirely before making a successful collision with a uranium nucleus.

The Core: 2. Water, which is (i) Moderator

For the purpose of slowing the neutrons, they are passed through a material containing small nuclei with which they can have elastic collisions without being absorbed and disabled.

Such a material is called a "moderator," a greatly misleading term. "Moderation," in ordinary language, means to tone down. The moderator in a nuclear reactor does indeed slow down the neutrons, but not for the purpose of making them less powerful. Indeed just the opposite is the purpose of this material: by slowing the neutrons, it makes them more potent as initiators of further fission. The moderator is there to *increase* the likelihood that any particular neutron will initiate a next-generation fission; it accomplishes this by slowing the neutron.

Two common materials used as moderators are water and graphite (carbon). Almost all reactors in the United States are water-moderated. Water has the advantage that it can double also as the carrier of the heat energy from the reactor core to the circulating systems that ultimately produce steam to drive the turbine of the electric generating plant.

The fuel rod assemblies are immersed in water. The neutrons given off in the fission process in one fuel rod assembly are too fast to produce much further fission in the same fuel rod, but after passing through the water, they can enter a neighboring fuel rod sufficiently slowed down to be potent fission initiators.

We have already discussed how the neutrons, originally going too fast to induce fission, are equilibrated to the moderator temperature. In millions of bouncy collisions they lose kinetic energy, and heat the moderator. The "hot" (fast) neutrons, colliding with the "cold" (slower) nuclei of the moderator, reach a common temperature in a few generations.

The diameter of the atomic nucleus is about 100,000 times smaller than the dimension of the atom, leaving vastly more empty space (occupied by electrons, which do not impede the movement of neutrons at all). It is apparent that these many collisions are not restricted to a small region, as might be the case in the mixing of hot and cold water, but spread to the entire volume of the core without the need for any "stirring" mechanism at all.

Neutrons that have reached the equilibrium temperature of the moderator are called, "*thermal neutrons.*" Thermal neutrons are not different from the other neutrons in any respect except their average kinetic energy. But the likelihood that they can initiate subsequent fission in uranium nuclei is much increased by their lower speed.

... and is also (ii) Primary Coolant

When the moderator is water, it can also perform the vital function of carrying away the heat energy that is produced by the fission of uranium. Because carrying away heat has the effect of cooling the core, the water is referred to as a "coolant." We will see later that the process of carrying heat from the core to the part of the nuclear plant that generates electricity, has to be done in two stages. The water in the core is in the first of the two loops that move heat energy, and is therefore called the "Primary Coolant."

The hot primary coolant water is pumped out of the core, and passed through a device in which heat is transferred into the secondary loop, leaving the primary coolant water cooler, ready to be pumped back into the core where it is heated again, and so on. In a continuous process, the primary coolant moves heat energy from the core to the secondary coolant, also water, which boils into steam to drive the turbine.

When the heat–moving function of the core water fails, for some reason such as a pump failure, suddenly the coolant function takes center stage. Not only does the system stop delivering energy to the electric generator, but the core begins to overheat. We will look at this situation in more detail when we discuss what can go wrong.

The Core: 3. Control Rods

If the core threatens to overheat, or some other mishap occurs, it is important to be able to shut down the reactor. For this reason and others, a set of rods is available to be lowered into the space between the fuel rod assemblies, to gobble up neutrons and disable the chain reaction.

These rods are connected together at the top, and move as a unit. They can be raised above the top of the core, or lowered all the way to the bottom, or held in any position between.

The control rods are made of material, such as cadmium or boron, whose nuclei absorb neutrons. Unlike the nuclei of the moderator, which share kinetic energy and then bounce the neutrons off, nuclei in the controls rods take the neutrons in and don't give them back. Neutrons that enter a control rod, do not come back out. They are dead as far as fission in the reactor is concerned.

The control rods perform the function of absorbing neutrons and "taking them out." in three different modes, serving three distinctly different purposes.

Fig 11.3 The control room, a fantasy by physics student Spencer Nuzum

Control Rods I: The SCRAM

There is a reason why the control rods are suspended from above. In fact, the structure that raises and lowers the rods is held at the top by an electromagnet. When this electromagnet lets go, the control rods literally drop down, and within less than a second, reach the bottom. Immediately they intercept all neutrons given off in fission in the entire core. When the control rods are all the way down, the generation ratio drops to zero, and all fission stops.

The electromagnet that holds the control rod assembly is powered from outside by an electric current that is controlled automatically. The electromagnet can also be manually controlled by operators in the control room.

In any one of a hundred different situations, the automatic control system shuts off the current to the electromagnet, the control rods drop down, and fission stops. This is an emergency measure called a "SCRAM," which stands for Safety Control Rod Adjustment Mechanism.

If a pump fails, if there is a leak in the reactor vessel, if the core overheats, if other control systems fail, there is a SCRAM. Since the entire reactor is controlled by electric circuitry, it is important that if the power goes down, the reactor stop generating heat. For this reason, the SCRAM is activated by the *lack of current* to the electromagnet. If all power is off, the magnet lets go.

A SCRAM is not a dire event, necessarily. It is quite normal for a reactor to experience several SCRAM's each year. A pump fails, a leak is detected, or some other rather ordinary event occurs, and for safety reasons the reactor is shut down. The pump is replaced, the leak is fixed, and the reactor is started up again.

We will see later that, while a SCRAM shuts down fission immediately, it does not completely terminate production of heat in the core. A SCRAM is an immediate but incomplete remedy for the most serious event to threaten the reactor, which is an overheating core.

Control Rods II: Short term, "Shim" Adjustment

The circuitry of the reactor control systems continuously adjusts the height of the control rods. Only the portions of the fuel rods that are below the bottom of the control rods participate in fission, because only there can the neutrons from the reaction in one fuel rod reach a neighboring fuel rod to initiate further fission.

If the control rods are lowered, less uranium is in play, because more of it is subject to neutron capture in the control rods. This lowers the "cold" (20°) generation ratio, which lowers the Operating Temperature, the temperature at which the TCXR line crosses the horizontal axis of the Excess reactivity vs Temperature graph (Fig 10.1).

The slight lowering or raising of the control rods is an effective adjustment mechanism for regulating the temperature of the water that is pumped from the reactor to the primary coolant loop.

During hours of the day when the demand for electricity is low in the communities supplied by the reactor (such as 2-4 am), less steam is needed to drive the turbine, less heat needs to be supplied by the Primary Coolant system, and the reactor can work at a lower operating temperature. The control rods are lowered slightly.

During hours of the day when the demand for electricity is high (such as 5-9 pm), more steam is needed to drive the turbines, the reactor must work at a higher operating temperature. The control rods are raised slightly.

As with most controls, this is automatic, but the operators in the control room can manually override the automatic control.

This function of the control rods is called the short term adjustment, because it occurs continually all day and night every day. It is also referred to as a "shim" adjustment; shim is a word meaning a fine adjustment. "Shim stock," for example, is paper–thin bronze sheet used by auto mechanics to raise the position of cylinder heads on the cylinder block.

Control Rods III: Compensation for Long term fuel usage

When a reactor has just been newly fueled, and the fuel rods are full of fresh uranium from top to bottom, it would be foolish (and dangerous) to use the entire length of the fuel rods for fission. At startup, a small portion of the rods at the bottom is sufficient to sustain the chain reaction and produce all the heat needed. The initial amount of uranium in the fuel rods is sufficient to run the reactor for up to a year. Refueling is a major task, and is not undertaken at frequent intervals. The fuel must last a long time.

So, at first, the control rods are lowered far down in the core, intercepting neutrons throughout most of the core. After some weeks and months, the uranium in the initially exposed portion of the fuel rods has been largely "spent," (used up) leaving the unfissionable portion of the uranium (the ${}_{92}U^{238}$) and the fission fragments, the smaller nuclei such as Krypton and Barium.

So, as the fuel is used, over the long term, the average daily position of the control rods is raised to bring new and unused portions of the fuel rods into the reaction. This continues for the life of the fuel rods, until at the end the control rods are nearly above the top of the core. This use of the control rods is called the "long term adjustment," and its purpose is, over the months, to compensate for the usage of the uranium, and to continually expose new portions of the fuel to participate in the chain reaction.

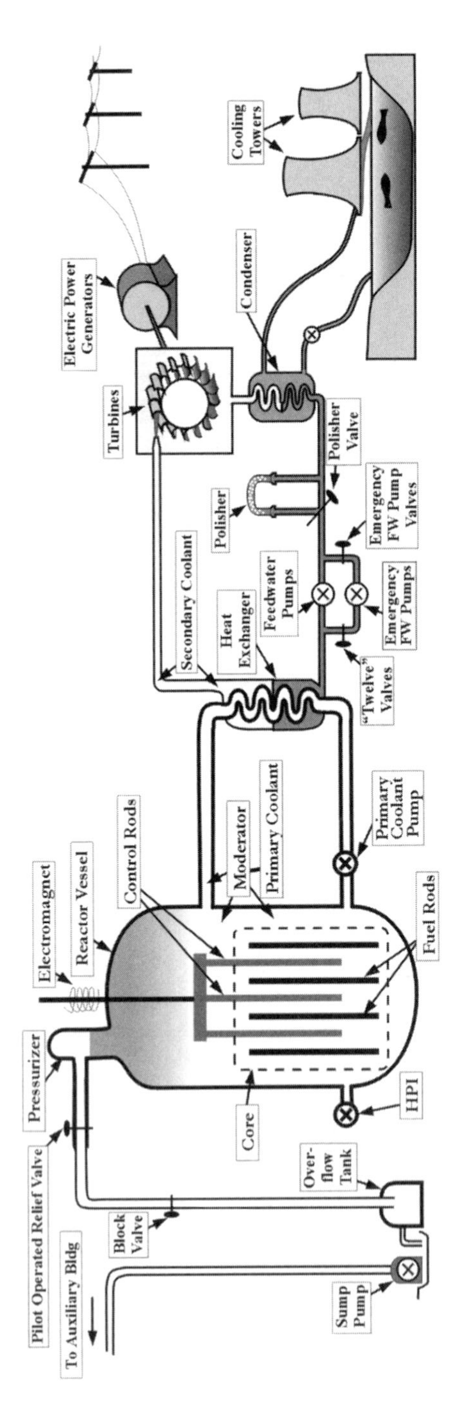

Fig 12.1 Schematic diagram of a Pressurized Water Uranium Reactor

12. Reactor Architecture II: Driving the Generator with nuclear heat

In the core of the reactor a controlled chain reaction produces a steady supply of heat, passed along mostly by the collision of fast neutrons with nuclei in the moderator, which is water.

The water in the core doubles as moderator and as primary coolant. As coolant, the water carries away heat produced in the core, sending it on the first lap of its journey to becoming electrical energy to heat homes, cool refrigerators, run motors, power electronic devices, and supply energy to industry, large and small.

In a coal, oil, or gas heated electrical power plant, the burning fuel boils water into steam that directly feeds the nozzles that drive the turbine that turns the electrical power generator. There are nuclear plants that boil core water directly for driving the turbine. These are called "boiling water reactors," or BWR's. But they are not common, for a very good reason.

The core water, although it is separated from the interior of the fuel rods by a hermetic seal, becomes slightly radioactive due to some leakage of gaseous fission products (such as krypton) and due to the steady presence of neutrons. Steam has a away of getting out. If this water were boiled and the radioactive steam released into the turbine, greater care would have to be taken that the turbine enclosure as well as the building in which the turbine is located are tightly sealed.

The secondary coolant loop

To avoid this problem, the heated core water, the primary coolant, is used to heat water in a secondary loop, using a heat exchanger in which heat is passed from core water to the secondary coolant, while the radioactivity remains sealed within the primary coolant system.

The primary coolant water, being under high pressure, can reach temperatures of several hundred degrees (Celsius) without boiling. This very hot water is passed downward through a coiled metallic tube which is surrounded by a "jacket" through which the secondary water passes upward. The metallic tube shields the secondary water from the radioactivity of the primary coolant, yet transmits the heat energy. In this exchange, the primary coolant is cooled and sent back to the core to be re-heated, while the secondary coolant is heated. The secondary water boils because it is not under the same high pressure as the core water. The device described is called the "heat exchanger," but it is also referred to as the "steam generator." The two names are interchangeable, and refer to the same device. The reader may be familiar with small heat exchangers, made of glass, that are used regularly in chemistry laboratories as distillation apparatus.

Reactors that generate steam in this two-stage process are called, "pressurized water reactors," or PWR's. Because the turbine steam is not radioactive in these reactors, the turbine can be away from the reactor, in a separate building connected to the containment building only by heat-insulated steam pipes. This alleviates any concerns about release of even small amounts of radioactive steam into the environment in the normal course of transferring heat from the core to the generator.

Fig 12.1 is a schematic diagram of the entire reactor complex, from the core to the wires that carry away the electric current at high voltage. The diagram is schematic. The objects in it are not drawn to scale, nor are the connections and placement necessarily as drawn.

The turbine and beyond

The steam generated in the heat exchanger passes through a set of pipes, and is released under pressure through nozzles aimed at the blades of a turbine. A turbine may be thought of as a large paddle wheel, like the turbines that are driven by falling water at hydro-electric power plants.

The turning turbine has a common shaft with the electrical generator. It is outside the scope of this book to explain how the conventional generator produces electrical energy.

Although it is not radioactive, the steam is recycled rather than allowed to escape into the atmosphere after having given the turbine blades the appropriate push. Because the secondary cooling system carries vast quantities of water, the amount of deposition of calcified minerals and other contaminants would be intolerable if the water were not purified. This is one of the reasons that the water is collected from the turbine, and used over and over. For thermodynamic reasons, the water must be cooled before it is once again pumped into the bottom of the steam generator.

The used turbine steam is cooled in another heat exchanger, designed somewhat differently, but along the same principles. Now the central coiled tube contains the steam after it has done its work in the turbine, while huge quantities of lake or river water are passed through the jacket to condense the steam. This heat exchanger is called a "condenser," because its function is to condense the used steam.

It is for this reason that nuclear reactors are almost always situated on the shore of a lake, or river. The cooling water that is passed through the condenser is filtered, but is otherwise not purified. This water usually emerges from 20 to 40 degrees warmer. Because there is so much of it, the region of the body of water where it is dumped could become warmer by several degrees, year round. Such a warming of the natural environment could be devastating for the water ecology in that area.

To bring this warmed water back to environmental temperatures before dumping it back into the lake or river is the function of the giant structures that dominate the landscape of the normal reactor site. These structure are called, "cooling towers." They are hollow structures of concrete, open at the top. The warmed water is brought in from the condensers at the bottom, and sprayed upward and outward onto the inside surface of the concrete towers, where it trickles down into a collecting pond. The giant towers themselves remain close to the temperature of the environment. By running down the tower walls, the water becomes equilibrated to environmental temperature before it is allowed to run back to the natural source.

The water that is sprayed up the interior of the tower is warm, and some of it evaporates, making up the clouds of water vapor that are usually seen emanating from the cooling towers. These menacing clouds are nothing but water vapor, and they dissolve in the air with no harmful effect; nevertheless, because they are commonly mistaken for smoke (nuclear smoke at that!), they are cause for more alarm than they deserve on the part of drivers seeing the cooling towers from a road that passes nearby.

Some minor components that became important

Pumps and a few other features designed for safety are shown in the schematic diagram in Fig 12.1 primarily because they played a major role in the unfolding of the 1979 accident at Three Mile

Island, in eastern Pennsylvania. An unusually readable minute by minute account of this accident is reprinted in the Appendix of this book. Some of the key pieces of equipment cited in the account are pictured in the schematic of Fig 12.1 to make the events described easier to follow.

The pumping of the primary and secondary coolant in their respective loops is one of the most critical functions of the reactor. If the pumping stops in either loop, there is a break in the chain of heat transmission from the reactor to the generating station that sends that energy to the consumer. Causing the generator to stop producing electricity creates a minor inconvenience. The more serious consequence of an interruption in pumping is that the heat ceases to be removed from the reactor core, causing a rapid and dangerous increase in temperature and pressure there.

Of course, a SCRAM is automatically triggered, stopping fission in less than a second. The SCRAM reduces, but does not stop the production of heat. Even though all fission has ceased, the radioactive fission products that remain in the fuel rods after fission continue to produce a sizeable amount of heat by radioactive decay, and there is no device that can stop, or even slow, that process.

For these reasons, enormous attention is given in the design of reactors to assuring that heat continues to be pumped away from the reactor, no matter what. Shown in the diagram are the pumps that normally move the primary and secondary coolant; emergency pumps that take over for these pumps should they fail; and pumps called High Pressure Injection pumps that replace primary coolant in case there is a loss of water in the core.

A standard engineering symbol for a pump is a cross =⊗= in a circle. Each such symbol in Fig 12.1 actually represents several pumps, because reserve pumping capacity should be immediately available in the event of the failure of a single pump. In addition, each set of pumps has a backup system of emergency pumps that can be turned on, automatically, or by the control room operators, to keep the pumping going.

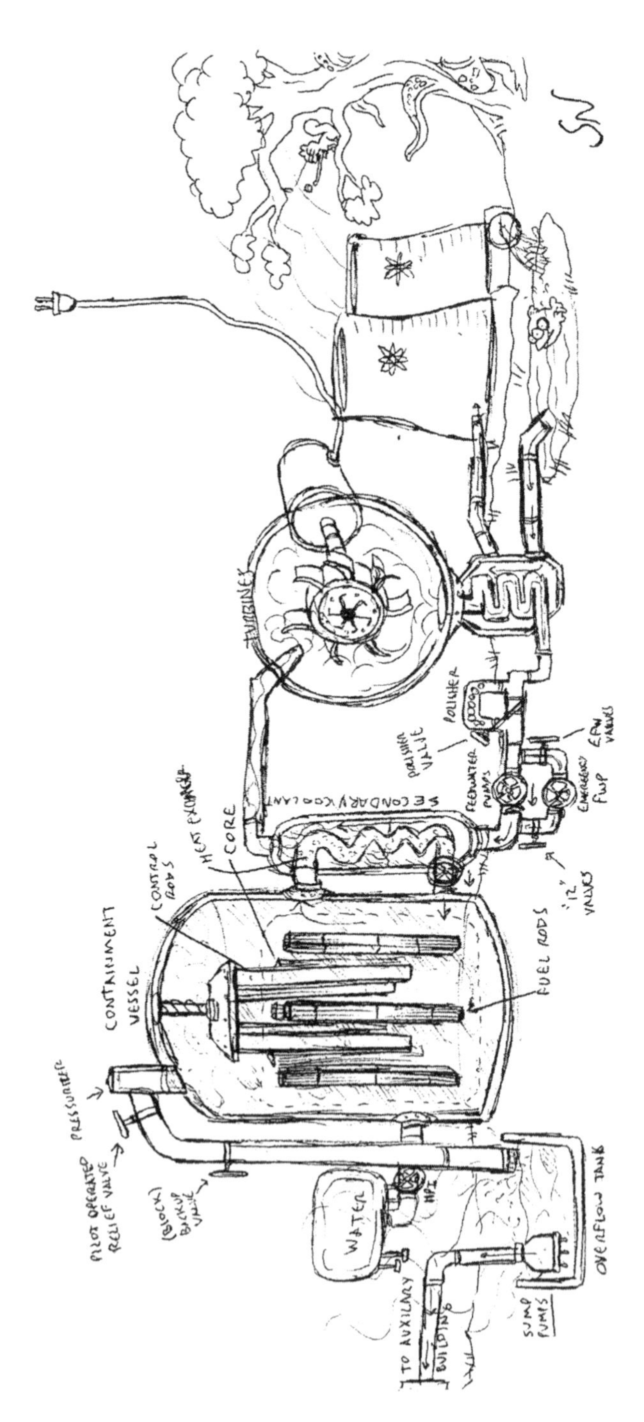

Fig 12.2 The reactor, a classroom sketch by physics student Spencer Nuzum

Familiarity with the technical jargon used in naming some of the features labeled in Fig 12.1 will later be useful in reading the account in the Appendix. The pumps in the primary coolant loop are called, naturally, "Primary Coolant Pumps." But for obscure reasons, the pumps in the secondary coolant loop are not called, "Secondary Coolant Pumps," but are called, "Feedwater Pumps."

If the feedwater pumps fail, there is a set of Emergency Feedwater Pumps that can take over. The branch in the plumbing that allows water to flow through the Emergency FW pumps is normally blocked by the Emergency FW Valves. When the Emergency FW pumps go on, those valves open automatically.

So that the Emergency FW pumps can be isolated from water under pressure on both sides, for servicing, a second set of valves, with the unlikely name, "Twelve valves," (there are not that many valves, that's just their name) are in the branch opposite the Emergency FW valves. The Twelve valves are normally open, but are shut when the pumps are being serviced.

The High Pressure Injection Pumps are heavy duty pumps that are able to pump water from a large reservoir tank at atmospheric pressure against the huge pressure difference into the reactor vessel. These specially designed pumps are shown in Fig 12.1 as "HPI."

A sump pump is also shown, which functions when water has leaked or overflowed onto the floor of the containment building and into a sump (a cavity built into the floor). From there, the water, which is slightly radioactive because it came from the core, is removed to emergency storage in an Auxiliary Building, which is normally used as a storage and work area.

A critically important control device is the "Pilot Operated Relief Valve" (the PORV) which is in the line from the pressurizer at the top of the reactor to the overflow system. This valve is normally closed, because the reactor water (the primary coolant) is under high pressure and must have no escape opening. If the pressure of the water in the reactor becomes dangerously high, the

PORV can be opened briefly to provide necessary relief by letting some water flow down the overflow outlet to an overflow tank on the containment building floor.

Both the water level and the pressure in the pressurizer are continuously monitored. When the pressure in the core exceeds a certain limit, the PORV is automatically opened, and then closed again when the pressure has returned to an acceptable level. If the automatic controls to the PORV fail, the valve can be opened or closed by control room operators. The values of water level and pressure are continuously displayed in the control room, as are indicators that tell the operators what valves are open or closed.

A second valve, called the "Block Valve" downstream in the overflow pipe from the PORV, is normally open, so that if the PORV is opened, the water can flow down the pipe. This valve is there so that if the PORV does not shut, an operator can close the Block Valve to prevent the loss of reactor water.

Finally, Fig 12.1 shows a bypass to the flow of secondary coolant water, that leads some of the water through a device called a "Polisher." The word "polish" means to shine up, make clean. The polishers are water pipes that contain an ion exchange resin, which works like many home water softeners, to remove dissolved minerals from the water. Because the secondary water over time accumulates minerals that would tend to deposit on the inside surfaces of the pipes, pumps, and in the heat exchanger, some of the circulating water is regularly passed through these resins to "polish," or clean the water. Polisher *valves* can open or close the water path through the polishers.

PART III

Some Other Reactor Technologies

13. Other coolant, other moderator

One variation of the Pressurized Water Reactor is a design used in Canada in which heavy water is used as moderator. Heavy water is H_2O made with the two-nucleon isotope of hydrogen, $_1H^2$, known as deuterium. We will encounter this important isotope in another use later. Deuterium occurs naturally with an abundance of about 0.015%. Although this is only one part in about 7000 atoms of hydrogen, considering all the water in the oceans, deuterium is plentiful. Because the mass ratio is two to one, separation of the "heavy" from the "light" hydrogen is not all that difficult or expensive.

Heavy water, moreover, is harmless. It is not radioactive, can be tasted and swallowed, or used for washing one's hands. Since it is chemically identical to ordinary water, it will go into recipes just as ordinary water does. The heavy water molecule has a molecular mass of 20 g/mole, as opposed to 18 g/mole for ordinary water.

The neutron darting about after being emitted in uranium fission, however, notices the difference, because in collisions with heavy hydrogen it encounters a partner twice its own mass (a proton plus a neutron) rather than an equal mass partner. The result is that heavy water is less absorbent of neutrons than ordinary, light water. Absorption of neutrons is a small and marginally undesirable property of ordinary water when it is used as coolant and as moderator. One would prefer the moderator to slow down the neutrons, without absorbing any.

There are several options in the design of heavy water reactors. Heavy water can be used directly in the place of ordinary water in the reactor described in Part II. A more popular design uses an elaborate arrangement so that the heavy water is used as moderator and remains in tubules in the core around the fuel rods, while a separate set of channels circulates ordinary water as coolant.

The fact that heavy water does not absorb neutrons means that the potential generation ratio is higher than when light water is used. It means that the level of enrichment with fissionable U^{235}, which is around 3% in light water reactors, need not be as high. In fact, it is possible to use un-enriched uranium, with only 0.7% of U^{235}, in heavy water reactors. The choice between heavy and light water tends to be an economic trade-off. With the maturing of the technology of enrichment, the cost of enriched uranium has decreased, and the economic advantage of the heavy water designs is no longer clear.

Graphite

Graphite has some advantages and some disadvantages as a moderator. Like heavy water, graphite moderates without absorbing neutrons. That it can maintain a chain reaction in un-enriched uranium was one of its chief attractions. The fact that graphite does not boil or leak has some advantages. But, of course, it liquefies at much too high a temperature to be able to double as a coolant.

Graphite moderated reactors have channels for the circulation of water as a coolant. These designs were standard in the Soviet Union.

At high temperatures the temperature coefficient of excess reactivity can turn positive in a graphite moderated reactor, at which point nature's nuclear thermostat starts to turn the heat up rather than down when it's too hot. This was one reason why the reactor at Chernobyl, in Ukraine, went out of control in 1986. The finger pointed immediately at the use of graphite as the culprit in the explosion of the reactor. The particular events that culminated in that disaster would not have happened in a water moderated reactor, but that is different from saying that water moderated reactors are safer. Certainly the reversal of sign in TCXR and the fact that graphite is combustible when it is overheated had an influence on the details of the explosion at Chernobyl.

Fig 13.1 Chernobyl reactor after it exploded. Was graphite moderator the culprit? (Soviet press agency Tass photo)

The sequence of unforgivable mistakes that brought that reactor to the point where it went over the brink is now well documented. Whether and how mistakes can be avoided is a question for later in this book.

14. The Plutonium reactor

Plutonium is element 94 in the periodic table. It contains two more protons than does Uranium, which has 92. Plutonium is not a naturally occurring element, because it is radioactive, and dissociates spontaneously, self-activated, by the emission of an alpha particle (2 neutrons and 2 protons). The radioactive decay of plutonium occurs with a half-life of 24,100 years. This means that any plutonium that might have been produced in the star that, by exploding as a supernova 4.5 billion years ago, gave us our rich array of elements, would have dissociated completely by now.

In addition to being radioactive, Plutonium is fissionable, and is used as a nuclear fuel in bombs and in reactors. Although it does not occur naturally, Plutonium is made artificially, out of the 99.3% of natural uranium that is unfissionable, and from the point of view of nuclear fission, is waste to begin with.

U^{238} is not fissionable; only the 0.7% of naturally occurring uranium that is U^{235} is fissionable.

U^{238} can not be made to split. It largely ignores the fast neutrons that come directly from the fission process, and it does likewise to thermal neutrons. However, over a range of speeds somewhat greater than thermal, there are narrow bands of neutron energies that interact with resonances in U^{238} so that these neutrons are captured by the Uranium-238 nucleus. This is a non-fissioning capture.

In the uranium fission reactor, measures are taken to prevent neutrons with speeds in that range from encountering the U^{238}, to prevent their capture. This is why the uranium fuel is placed in the core in the form of pellets, surrounded by large spaces filled with moderator. In the process of equilibrating their kinetic energy with that of the moderator, the neutrons invariably have to pass through that range. The only way to prevent their capture by U^{238} is to make sure the neutrons stay in the moderator, away from all uranium, until they have completely passed through that range.

This separation of fuel and moderator in a "lumpy" combination is crucial to the design of the standard uranium reactor.

And if it is not?

If, instead, the uranium is pulverized and mixed intimately with graphite or water, the neutrons will continually dart back and forth between moderator and uranium dust. During the critical part of the equilibration when they are not quite thermal, and are in the range where they can be absorbed by U^{238}, many would be.

When neutrons are absorbed by uranium-238, not only are they lost to the uranium chain reaction, but the nuclei formed from the capture begin a sequence of events that ends with plutonium.

If it is desired to make plutonium, the pulverization and intimate mixing of uranium with moderator is done deliberately. Here is what happens.

When a nucleus of U^{238} absorbs one of the near-thermal neutrons, it becomes U^{239}.

U^{239} is not stable. It dissociates over about a half hour by emitting a beta particle (written, β), which is an electron. (Radioactive dissociation will be discussed in section IV.) The electron that is emitted in radioactive decay has its origin in the nucleus, and is not an orbital, or chemical, electron. A neutron *in the nucleus* breaks up into a proton and a high energy electron that is emitted from the nucleus with great energy.

The nuclear electron is one product of a neutron decay in the nucleus, that leaves behind a new proton in its place. The result of the beta emission from U^{239} is a nucleus that is short one neutron but has in its place a new proton. The new nucleus still has 239 nucleons, but it now has 93 instead of 92 protons. It is therefore no longer uranium, but the next element up the scale in the periodic table, an artificial element, called "Neptunium." This sequence of two steps is written as follows,

$$_{92}U^{238} + {}_{0}n^{1} \rightarrow {}_{92}U^{239} \rightarrow {}_{93}Np^{239} + \beta \qquad [15.01]$$

Neptunium, is also unstable, and dissociates in a few days by emitting another beta particle, leaving the nucleus once again richer by one proton but with the same total number of nucleons. It is now an element with the atomic number, 94, called, plutonium.[1]

$$_{93}Np^{239} \rightarrow {}_{94}Pu^{239} + \beta \qquad [15.02]$$

Plutonium is radioactive, but will keep for 20,000 years.[2] More important, it is fissionable.

And so, from the neutrons produced in the initial uranium chain reaction, come not only the next generation of fissions in uranium, but new fissionable nuclei as well. The net outcome is that the three neutrons from an original fission can initiate the one next-generation fission that is required to keep the chain reaction going, and in addition make out of non-fissionable U^{238} at least one new fissionable nucleus, a plutonium nucleus.

---

[1] Astute readers with some knowledge of the planets, will have observed that the naming of the three elements beginning with Uranium follows the names of the three outermost planets of the solar system: Uranus (uranium), Neptune (neptunium) and Pluto (plutonium).

[2] The half life of plutonium is 24,100 years. In that time, half of an original sample has dissociated.

Plutonium can be split in much the same way that U^{235} is split, by hitting it with a thermal neutron. The plutonium reactor must be started with an initial supply of fissionable material, be that U^{235} or plutonium. As that initial fissionable material is used up, it is replaced by newly made plutonium. The plutonium not only feeds the chain reaction that keeps it going, but replenishes itself until all the non-fissionable uranium (U^{238}) is used up. In effect it makes fissionable material out of stuff that originally was non-fissionable, and then consumes it as nuclear fuel.

The inexhaustible fuel (not really)

It seems an energy engineer's dream come true. It has been said that the plutonium reactor "makes more fuel than it uses." This conjures up the image of an automobile driven 200 miles ending up with more gasoline in the tank than it started with.[3]

Plutonium reactors have been called, "breeder reactors," suggesting that they "breed" new fuel. It is indeed economically very sound to make use of the 99.3% of naturally occurring uranium that is, as uranium, non-fissionable. Of course, unlike rabbits, breeder reactors cannot "breed" without limit. A plutonium reactor can "breed" new fuel only as long as its supply of U^{238} lasts.

European nations with nuclear programs, France notably, have standardized a particular design of plutonium facilities and built many of them, not only for domestic installation, but for export. Many countries without the resources to build their own, have bought breeder reactors. These reactors have a remarkably good

---

[3] An overenthusiastic author has exclaimed, "breeder reactors ... could produce all the energy mankind will ever need," while ordinary uranium reactors would deplete the world's nuclear fuel in 50 years. (Bernard L. Cohen, "Before It's Too Late - A scientist's case for nuclear energy," Plenum, NY, 1983)

safety record. They appear to be well designed, and standardization has limited the variability that comes with continually changing blueprints.

There is a major concern about these reactors. Plutonium, being a chemically different element from uranium, can be separated from the non-plutonium material in partially spent fuel rods by chemical means, almost 100% pure. This makes it possible to recover weapons grade fissionable material from a breeder reactor. There is no easy way to tell whether some of the plutonium in such a reactor has been skimmed for weapons production.

It is not possible to similarly recover weapons grade (90% enriched) uranium from an ordinary uranium reactor.

15. Fusion

The graph of Fig 3.2 shows the "energy remaining" in the nuclei of various isotopes. That graph leads us to uranium on the high end, which can be re-arranged in smaller, more tightly bonded, fragments, yielding energy from the reaction [5.01].

It leads us on the low end to the dramatically large energy releases in the joining, or "fusion," of small nuclei, and in particular to the reaction of deuterium with tritium to produce helium and a neutron. This reaction was described in chapter 4. Both Eq 4.09 and the diagram of the reaction are reproduced here.

$$_{1}H^{2} + {}_{1}H^{3} \rightarrow {}_{2}He^{4} + {}_{0}n^{1} + \text{Energy yield} \quad [4.09]$$

The energy yield of this reaction is 1.7×10^{12} Joules/mole.

This reaction is one of several that take place in the interior of the sun, providing the energy which we, on earth, receive from the sun.

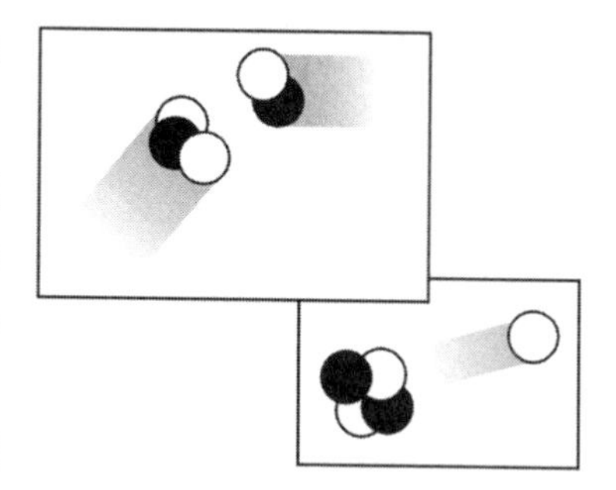

Fig 15.1 Hydrogen fusion

The reason that we have had limited success in reproducing this reaction on earth is the extreme nature of the activation that is needed. The obstacle to the fusion of these two nuclei is the mutual repulsion of the proton in each nucleus, which is enormous at nuclear distances. The only means known so far to provide the force to overcome this repulsion is a temperature of over 200,000,000 degrees. At that temperature, the average kinetic energy of the nuclei will occasionally drive them toward each other with enough speed so that the repulsion can not stop and reverse their path before they fuse.

Temperatures of that magnitude have been produced by exploding a uranium bomb into a mixture of deuterium and tritium. This is done in the thermonuclear warheads that are in the tips of most military missiles.

The "thermonuclear" bomb is a weapon of unparalleled destructive power. Where the city of Hiroshima was leveled by a uranium bomb with the destructive equivalent of fifteen thousand tons of TNT, thermonuclear bombs carry the equivalent of one million tons of TNT.

We haven't yet been able to make this reaction go under the controlled conditions necessary in a reactor, on an economically profitable scale.

It's not because they haven't tried

For over a half century, scientists and engineers have been seeking to bring nuclear fusion under control, so that it could produce energy at a moderate and sustained rate, as is required in a reactor. Nuclear fusion has many advantages over fission. Its end product is ordinary helium, which is neither radioactive nor toxic, and can just be released into the air. Its fuel consists of two components, deuterium which exists naturally and plentifully, and needs only to be separated from sea water, and tritium, which has to be made, but can be manufactured at reasonable cost. Tritium is radioactive, but can be stored relatively safely, decaying to He^3 with a half-life of 12 years.

It has not been found practical to use accelerated *aimed* deuterium and tritium nuclei, in colliding beams, as is common in experimental particle physics. The cost of accelerating them would be much greater than the yield in energy.

Confinement

Temperatures high enough to activate nuclear fusion of deuterium and tritium can be produced in laboratories by the use of lasers. The difficulty is that in the process of heating a mixture of deuterium and tritium to such temperatures, every conceivable material that might be used as a container to confine this mix would be vaporized. The result would be that the deuterium and tritium would simply scatter before fusing.

The problem is one of "confinement," finding a way to keep the reactants confined in a small enough region for the duration of time required by the fusion process.

The interior of the sun is kept compressed and confined by the pull of gravity on the upper layers of the sun that weigh down on the interior with immense force.

In a thermonuclear bomb, a conventional uranium or plutonium bomb is exploded as a trigger to the fusion reaction. The energy of the fission bomb is used to heat a deuterium-tritium mixture to fusion temperature. The explosion is so rapid, that the fusing nuclei do not have time to scatter before they undergo fusion. The inertia of the material, the sluggishness property that resists acceleration, holds it in place until fusion occurs. The name given to this strategy is, "inertial confinement."

Little glass beads

Inertial confinement consists of surprising the deuterium-tritium mix with very rapid heating, so that the fusion process wins the race against the dispersion of the material. Although achieving inertial confinement on a small scale has been difficult, it has been one of two major strategies that have been pursued over the past half century as a means of making nuclear fusion possible in a reactor.

Inertial confinement has been achieved on a small scale by filling tiny hollow glass beads with a deuterium-tritium mixture, and blasting them with a quick pulse of a high energy laser beam sufficient to produce the necessary temperature. Fusion of deuterium and tritium has been accomplished by this method, but not yet on a large enough scale to be practical. In a reactor, a continuous succession of such reactant-filled glass beads would have to be thrust into the target area, and blasted there by the laser. The critical break-even point would be reached when the reward of energy from the fusion can exceed the cost of staging such a scene on an ongoing basis.

The Tokamak

The other major attack on reactor-scale fusion activation has been the development of "magnetic confinement." A device called a "Tokamak" consists of huge magnets that focus a carefully patterned magnetic field in a region where the fusion is to take place. In a strong magnetic field, charged particles move in circles (in helices, to be more precise), and it is the idea here to see whether the charged positive nuclei (tritium and deuterium) can be made to orbit in such tight circles that they are "orbitally" confined by the magnetic field. Tokamak fusion has been achieved in experimental devices, over many generations of steadily improving apparatus. The measure of success has been the degree to which the process can be sustained with a net gain of energy. There is optimism that in time such an arrangement can produce net energy yield over the long duration.

There are always other ideas coming forward. One of these, although it is referred to as a fusion device, makes use of a fission reaction in the left part of the "remaining binding energy" graph. The graph zig zags in the portion between atomic mass numbers 3 and 12 into a deep dip at Helium (A=4), with ordinary Boron, ${}_5B^{11}$, (A=11) having higher "binding energy remaining" per nucleon than Helium. By splitting a boron nucleus into three helium nuclei, there is actually a net energy release. It turns out that boron can be split by hitting it with a proton, in the following reaction,

$$ {}_1p^1 + {}_5B^{11} \rightarrow 3\,({}_2He^4) + \text{Energy} \qquad [15.03] $$

The proposal is to have this happen in colliding beams of protons, electrons, and boron nuclei.[1]

Can we afford unlimited cheap energy?

In the race for cheap energy, there is a question that has been largely ignored. Albert Bartlett called attention to it recently.[2] It is an environmental question.

If, perchance, we can make nuclear fusion give us energy "too cheap to meter," continued growth in energy consumption from an inexhaustible fuel source, the water in the ocean, would we heat our planet so much that we would be unable to survive? Bartlett calculates that "in an estimated 13 doubling times of energy consumption, the numbers suggest that consumption by the human race would equal the sun's input to the earth." This would throw off

---

[1] Rostoker, et al, "Colliding Bean Fusion Reactor," *Science* **278** pp 1419-22 1997.

[2] A.A. Bartlett, "Fusion and the Future," *Physics and Society* **18** #3 1989 pp11-12.

the balance between incoming and outgoing heat energy that keeps the earth at a moderate temperature.

Nuclear fusion has the desirable feature that its reaction products are harmless. Helium is chemically inert and is not radioactive. What is ignored in this scenario is the other kind of reaction product, energy.

No matter what we do with the energy that we use, whether we produce heat directly, as in toasting bread, or drive electric motors and pumps and refrigerators, or light our homes, or run our electronic devices, it is a law of thermodynamics that it all tends to end up as heat. Quite apart from the greenhouse effect, Bartlett suggests that just from the point of view of the heat we would generate if electric energy truly became virtually free and unlimited in availability, we could not afford to use it without restraint.

PART IV

Radioactivity

16. Radioactivity: What it is

Radioactivity can be described in cold, scientific terms, but like no other aspect of the subject of nuclear energy, it evokes feelings of fear and apprehension. That the effects of radioactivity are delivered by stealthy, invisible, silent forces only intensifies these feelings.

A farmer whose family raised sheep and cattle near the Hanford Reactor in eastern Washington State drives his guest around the farm where he grew up and still now lives.[1] He points to a fence, and says, "We didn't think about it much. Their business was reactors, and ours was farming. They told us it was safe. That's what they told us."

Then, he goes on, they noticed the birth of increasing numbers of deformed animals. Animals who couldn't stand, who couldn't see, who died. And they began to wonder, "If this is what's happening to the animals, what about us?" He drives the visitor in his truck along a stretch of farm road that he says has been called the "Death Mile." As they pass homes, he points, and says, "Cancer, there." "Two children were born deformed, one blind." He adds that he himself has a genetic disability.

Deformed animals were the basis of a class action law suit for damages sustained by the "downwinders" of Nevada and Utah, the residents downwind from the atomic bomb tests in the 1950's.

Compensation for damages have been awarded to nuclear plant workers in Fernald, Ohio.

---

[1] Academy Award winning "Best Documtary," *Deadly Deception*, produced by Women's Educational Media, Inc. in 1986

Karen Silkwood, a worker in a plutonium reprocessing plant in Oklahoma died when her car ran off the road while she was on her way to meet with a New York Times reporter to tell her story of safety violations in the plant, and of her suspicion that she had been intentionally contaminated with plutonium after she protested conditions. Her estate won a large settlement from her employer. Millions know her story from the movie, "Silkwood," in which Meryl Streep plays Karen Silkwood.

The cleanup of nuclear waste carelessly dumped around nuclear facilities has just begun. The companies running many such facilities perpetrated acts of carelessness and total disdain for the safety of the public, under immunity from federal regulation bestowed upon them through government contracts on the grounds of national security. It will cost billions, and take upwards of twenty years, to clean up.

There is widespread suspicion and mistrust of the assurances that are frequently given by the nuclear industry and those who contract with it that *now* we have enough safeguards in place, and that *now* all harm and danger are in the past.

Statistical evidence researched by epidemiologists and other scientists who have studied the population effects of radiation exposure indicate that thousands of people have suffered injury, given birth to deformed children, or died prematurely of cancer. It is almost never possible to assign the cause for any one person's cancer or genetic damage. This has limited the success of legal remedies for individuals. Being one in a population that has sustained undeniable statistical injury is little comfort.

And so we take a serious look at radioactivity, what it is, how it is measured, and how the average citizen can best evaluate information it receives from science, industry, government, and advocacy organizations.

What is radioactivity?

Radioactivity is the spontaneous dissociation of an unstable nucleus characterized by the emission of one or more of four particles, identified by Greek-letter names. Fission is usually not included in this category.

The principal radioactive particles

Alpha The alpha particle (α) is a very stable little package of two protons and two neutrons, and is in fact a helium nucleus ($_2He^4$). This particle is not highly penetrating, but, where it does reach, is the most damaging to living matter.

Beta There are two beta particles, one negative and one positive. The most common is the negative beta (β^-) It is an electron, produced in the nucleus by the separation of a neutron into a proton and an electron. This electron is not one of the orbital electrons that make up the chemistry of the atom. Its ability to do damage comes from its extremely high speed. The beta particle is moderately penetrating. When it is emitted from radioactive atoms that have become lodged in the internal organs of the body, it can do damage there.

Positive "electrons," are anti-matter, and are called positrons (β^+). Positrons do not long survive in a world of ordinary matter, usually succumbing by mutual annihilation with an ordinary electron, a process that emits a high energy light particle.

Gamma The gamma particle (γ) is a high energy "wave packet" of light. It is electromagnetic radiation. Its high energy and the absence of electric charge make it extremely penetrating, able to pass through to any portion of a human body. When it interacts with an atom, a gamma particle is able to knock electrons out of their orbit, ionizing the atom or molecule, with the possibility of producing chemical alteration.

Radioactive dissociation is a transition that allows the nucleus to "roll down the hill" to a lower and (from its perspective) preferable energy state. The energy is released, principally in the emitted particle.

Radioactive materials are in general not suitable as fuels, because the release of energy is self-activated, and we can not control its timing. The exception is the unusual case in which there is a steady, predictable, constant demand for small amounts of energy over a long time. An example of this is the fueling of an electric power cell on a space craft's journey to a far distant place, such as the outer planets and beyond. On such a journey, the energy to power radio transmission back to earth, and to receive signals from earth, can be supplied by plutonium used as a *radioactive* source (*not* a plutonium *reactor*).

The question then arises, what causes these radioactive materials to remain in a higher energy state, and undergo these spontaneous transitions that release their excess energy, one atom at a time, at some internally controlled rate. Some escape their unstable high-energy state through radioactive decay very quickly, some take thousands and millions of years.

The timing of radioactive dissociation

The question of timing is a very important one. Suppose that a nucleus is unstable, and could go to a more stable configuration by undergoing a radioactive dissociation. When will it do that?

The reason it does not do it immediately is that there is an energy barrier, a temporary state of higher energy that must be hurdled. Imagine a collection of bouncing balls in an open box on a table. If one of these balls can get enough energy to jump over the rim of the box, it can then fall to the floor, releasing the energy of height with which it was trapped inside the box.

On the average, the balls do not have enough energy to jump the rim. In their random bouncing about, the balls exchange kinetic energy. Some temporarily lose energy in collisions with other balls, some temporarily gain energy. If we look in on this bouncing around for a while, there is a random chance that at some time, maybe sooner, maybe later, we will see one ball go over the rim, and roll down.[2]

For any such arrangement of balls in a box, with whatever bouncing energy is distributed among them, there is a statistical likelihood that in any specified period of time, let us say a second, one will go over the rim. And so if we observed a number of identical boxes with identical balls of identical average bouncing energy, there would be a characteristic average time when one ball would surmount the rim. We could wait until half of these identical boxes had lost a ball. If the number of boxes that we observe is large, then the statistics would predict rather well when half would have lost a ball. Statistics apply to large samples, where individual variations in a random process tend to average out.

The time when half the boxes would be expected to have lost a ball is called a "half life."

There is a similar process that goes on in the unstable nuclei of radioactive isotopes. The nucleons (protons and neutrons) in a nucleus form a configuration that behaves more like a liquid droplet than like a solid mass. And so one can visualize the nucleus as a liquid drop that is undulating with its own energy, deforming first with a dimple here, then with a little flat spot there, then with a bulge there, and so on. The emission of an alpha particle can be visualized as resulting from a chance deformation that produces a pointy bulge, in which this very stable little package of two neutrons and two protons has a chance to separate itself somewhat

---

[2] In this analogy, which is not perfect, it is assumed that the loss of one ball represents the emission of one radioactive particle, and that once one ball is gone, the ones remaining are "satisfied," and no longer seek to escape.

from the main liquid drop. When it does so, because it is positively charged, as is the rest of the nucleus, there is suddenly a tremendous electrostatic repulsion, that overcomes the more distance-dependent nuclear force attraction, and off goes the little alpha particle, shot as if from a gun.

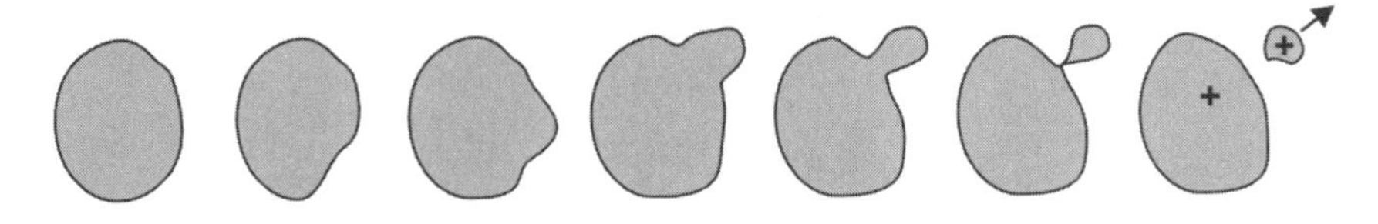

Fig 16.1 Liquid drop model of a nucleus. An alpha particle pinches off and is emitted in a radioactive decay.

Other, less easily visualized, chance variations are propitious for the processes that cause a neutron in a nucleus to suddenly decay into a proton and a high-energy electron. The emission of a gamma particle often accompanies other dissociations, and is analogous to the relaxation of an excited state of an orbital electron that accompanies phosphorescent light emission.

Half life

In radioactive emissions, random fluctuations occasionally create the conditions under which a nucleus can overcome an energy barrier to a particular dissociation. Random means that one can not know when it will occur, except that one can know what the likelihood is that it will occur within a given interval of time. For each radioactive nucleus there is a characteristic half-life, during which half of any collection of these nuclei will have dissociated. We designate this half life with the symbol, $t_{1/2}$.

What is the "half life" a half of?

It is not true that if you wait two times a half life, all the nuclei will have dissociated. The half life is the time during which half of any population of radioactive nuclei will have dissociated. In the second half life, only the half that still remain radioactive will dissociate, meaning a number equal to just one fourth of the original number will dissociate, leaving one quarter of the original number still radioactive. In the third half life, one half of those remaining will again dissociate; this time that will be 1/8 of the original number, leaving 1/8 still radioactive. Those that have dissociated at each stage have left the game, and no longer count.

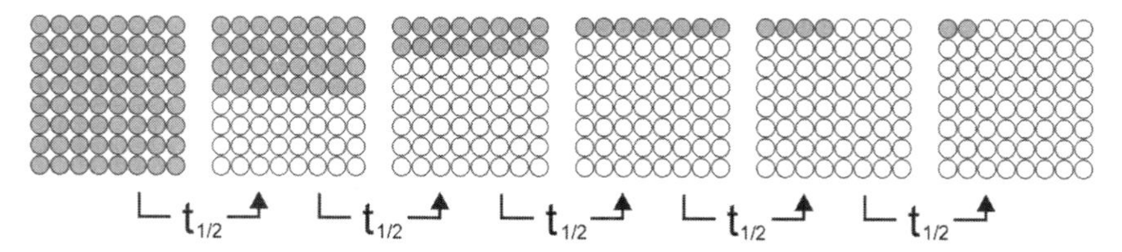

Fig 16.2 Radioactive decay. Shaded atoms are radioactive; open circles represent atoms that have dissociated by emission, and are no longer radioactive. In each half life, half of those still radioactive will dissociate.

32 children and their hats

There is a little story to help get this point across. Suppose that you take a group of 32 first graders on a field trip to a baseball game. The home team is, let us say, the Detroit Tigers. On the return home each child is given a baseball cap with the Tigers emblem. The following day, all children appear in school wearing their Tiger hats.

First graders tend to lose things like hats, and the statistics show that it is likely that in the first two days, half the children will lose their hats. So, after two days, only 16 children still have hats. Those who have lost their hats no longer play in this game; only those who still have hats. Statistics then predict that in the next two days half of these 16 will lose their hats, and only 8 will still have their hats. In another two days, half of these 8 will lose their hats, and only 4 will have hats. Two days later only 2 will have hats. And two more days later only 1 will have a hat. In the next two days, there is a 50% chance that this one will lose her hat. The laws of statistics apply in a different way to such small numbers. In fact, the laws of statistics have long since stopped applying exactly to the numbers of children; it is quite possible that in the second two days, 9 will have lost hats, or 7, or 6. But when the numbers are large, which they are when we are dealing with radioactive nuclei, statistics apply much more exactly.

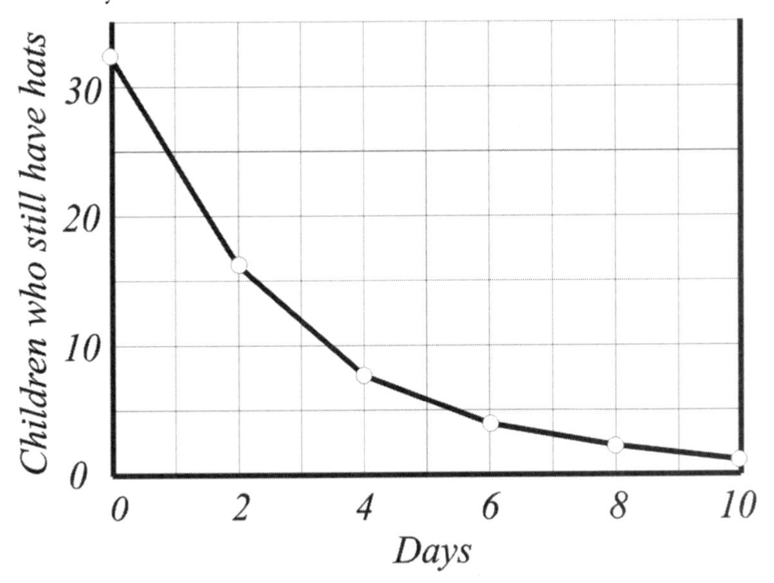

Fig 16.3 Number of children who still have hats. Every two days, each child who still has a hat has a 50% probability of losing it. The half life of hats is 2 days.

Problems

16.1 Can you think of three other things in ordinary life that disappear at a rate proportional to how many are left?

16.2 No guest wants to eat the last slice of a delicious pie, and so when there is just one slice left, those who want to have just a little more divide whatever is left into two smaller slices, taking one and leaving one. Starting with one full sized slice, write what fraction of a full sized slice is left after each of the seven guests has helped themselves to just a little more. Then graph your data.

17.
How radioactive?

It might be thought that if in the first two days 16 children lose their hats, 8 of those hats will be lost on the first day and 8 the second day. But of course that is not so. At the beginning of the second day, there will be fewer children with hats than at the beginning of the first day, so it is likely that of the 16, more will lose their hats on the first than on the second day. Perhaps 9 on the first day and 7 on the second day? Or 10 and 6? How are we to tell?

It was probably not a good idea in the graph of Fig 16.3, to connect the points with straight lines. Doing so neglects the fact that as soon as some hats are lost the rate of hat-losing goes down. It is more accurate to curve the line to go smoothly through the points. In Fig 17.1 this is done. Now, the curve begins to resemble, in its shape, the exponential decrease curve of Figs 9.1 – 9.4. That is not a coincidence. Radioactive dissociation is an exponential process.

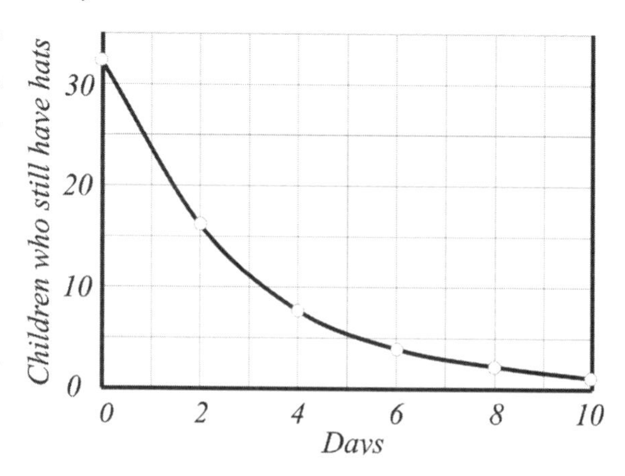

Fig 17.1 Points in the data of Fig 16.3 connected with a smooth line. Losing hats is a continuous process whose rate is always tied to how many children still have theirs.

Time constant, rate constant

Indeed, the graph of Fig 17.1 is an exponential decrease curve. Exponential functions describe situations in which a *rate* of change in the number of objects is just proportional to the number of objects contributing to the possibility of such a change.

Let us examine what this means for radioactive dissociation. Let us assume that we begin with a number, n_0 of radioactive nuclei. In the next small interval of time, Δt, each nucleus contributes a tiny probability, p, of its own dissociation to the total number of dissociations. This probability will surely be proportional to how long the time interval is, so p will be some constant times Δt. If this constant is called, λ, then

$$p = \lambda \, \Delta t \qquad [17.01]$$

Multiply this probability by the number, n_0, of radioactive nuclei that each contribute a probability, p, of dissociating, and the number of dissociations in Δt is $n_0 \times p$ or $n_0 \times (\lambda \, \Delta t)$.

The dissociations occurring in Δt will cause a small change, Δn, in the number of remaining radioactive nuclei.

$$\Delta n = n_0 \times \lambda \, \Delta t \qquad [17.02]$$

This allows us to write an expression for the *rate* of dissociation in n_0 radioactive nuclei. This *rate* is the number of dissociations *per second*, and is called the radio-activity, or just the "activity" (when the context is clear). The symbol for radio-activity is a,

$$a = \Delta n / \Delta t = n_0 \, \lambda \qquad [17.03]$$

The unit of radioactivity is *dissociations per second*. This unit has a name. One dissociation per second is called one Becquerel, after Henri Becquerel, who, along with Marie and Pierre Curie, are credited with the discovery of radioactivity.

Equation 17.03 can be solved to give the number of radioactive nuclei, n, that remain after a time, t (when the number you start with is n_0). The solution is the exponential function, $n = n_0\, e^{-\lambda t}$. This is a rather abstract function, in which e is a mathematical constant which is equal to approximately 2.718. This equation begins to make some sense when it is expressed in terms of a time constant called τ, which is the inverse of λ.

$$\tau = 1 / \lambda$$

What is the meaning of this time constant? No, it is not the half life, but it is related to the half life. Write the exponential function in the previous paragraph with $1/\tau$ in the place of λ, and it looks like this:

$$n = n_0 \; e^{-t/\tau} \qquad [17.04]$$

When an amount of time equal to the time constant has elapsed, the fraction t/τ is equal to one. At that special time, $n = n_0\, e^{-1}$. The number of radioactive nuclei has decreased to $1/e$ or of its original value. Where the half life is the time during which the number of radioactive nuclei decreases to half its original value, the time constant is the time during which that number has decreased to 1/2.718 of its original value. Whenever one time constant elapses, the number of radioactive nuclei declines by this same ratio.

This might not seem like progress. But look at the expression for radioactivity in eq 17.03. With $1/\tau$ in the place of λ, [17.03] becomes,

$$a = \Delta n / \Delta t = n_0 / \tau \qquad [17.05]$$

This means that if you start at a time when there are n_0 radioactive nuclei (which can be any time you choose), *and* if you could maintain the dissociation rate at its initial value, *then*, in an amount of time equal to τ, ***all*** n_0 nuclei that you started with would dissociate.

Be patient a little longer. Of course, you can't actually do this. There is no way to freeze the rate of dissociation; as soon as the number of radioactive nuclei starts to decrease, so does the dissociation rate. We have asked you to suppose what would happen *if* you could do something that you can't do.

Bring back the "Children Who Still have Hats" graph of Fig 17.1. We know what the half life is for those hats: It is two days. After two days, only 16 children have hats.

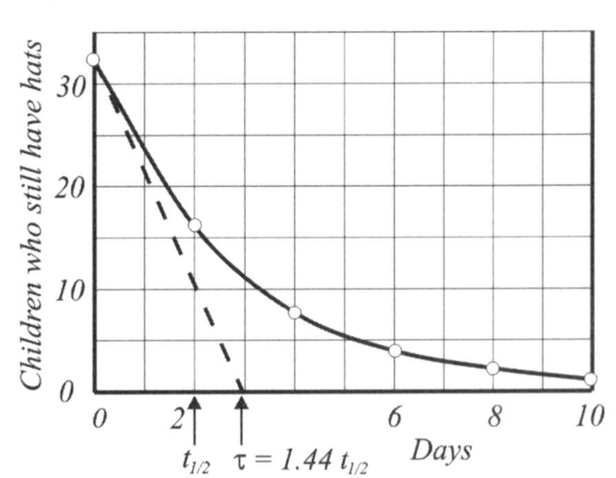

Fig 17.2 The dashed line tells what would happen *if* the initial rate were frozen fixed

A dashed straight line has been drawn tangent to the original line, at the beginning, before any hats were lost.

The dashed line represents what *would have* happened *if* the children had continued to lose hats at the original rate, which of course they didn't because there were ever fewer children with hats.

But, where the dashed line crosses the "Days" axis is the value of τ. That line crosses at 2.88 days. That value has a relation to the half life. It is 1.44 times the half life. τ is always, in every exponential decline function, equal to $1.44\, t_{1/2}$.

We have finally reached a valuable result. It allows us to re-write equation 17.05 as follows,

Equation to find radiactivity

$$a = n_0 / (1.44 \times t_{1/2}) \qquad [17.06]$$

Most tables of radioactive isotopes list the half life, not λ or τ. This makes eq 17.06 useful.

Remember units!

In using Eq 17.06, one must remember that n_0 is *the number of radioactive nuclei*, not the mass or the number of moles. $t_{1/2}$ must be *in seconds*, not in the units that may be attached to the half life in a table. Only then is the result in *dissociations per second*, or Becquerels

Finding radioactivity

==

Example

Find the radioactivity of 0.02 grams of the radioactive isotope of carbon, ${}_6C^{14}$. The half life of this isotope is 5730 years (see Table 17.1)

1. calculate n_0

The isotope mass is the mass of one mole of the isotope, in units of grams/mole. The isotope mass of carbon–14 is 14.00 grams/mole (*in these calculations, the integer in the superscript of the isotope symbol is close enough to use in place of the mass number*).

$$n_0 = \text{mass} \times (\text{number of atoms per gram})$$

$$n_0 = 0.02 \text{ grams} \times \{(1 \text{ mole}) / (14.00 \text{ grams})\}$$

$$n_0 = 0.02 \text{ grams} \times \{(6 \times 10^{23} \text{ atoms}) / (14.00 \text{ grams})\}$$

$$n_0 = 8.57 \times 10^{20} \text{ atoms}$$

2. Calculate the half life in seconds

$t_{1/2} = (5730\ \text{yr}) \times (365\ \text{da/yr}) \times (24\ \text{hr/da}) \times (3600\ \text{sec/hr})$

$t_{1/2} = 1.807 \times 10^{11}$ sec.

3. Calculate the radioactivity

$a = n_0 \,/\, (1.44 \times t_{1/2})$ [17.06]

$a = 3.294 \times 10^{9}$ dissociations per second or 3.294×10^{9} Bq

===

One dissociation per second is an almost imperceptible rate of radioactivity. The Becquerel is a very small unit. A more practical unit of radioactivity is one that was introduced before it was known what radioactivity is. It had been agreed that the radioactivity of one gram of radium will be called a *Curie* with the abbreviation, *Ci*.

One gram of radium has a dissociation rate of 3.7×10^{10} Bq. This provides the conversion factor.

Conversion between Curies and dissociations per second
1 Ci = 3.7×10^{10} Bq (disso/sec) [17.07]

With this conversion, the result obtained above for the radioactivity of 0.02 grams of carbon–14 is,

$$a = 3.294 \times 10^{9}\ \text{Bq} \times 1\ \text{Ci} \,/\, (3.7 \times 10^{10}\ \text{Bq}) = 0.0890\ \text{Ci}$$

When is NOW?

Eq 17.06 tells us that the *radioactivity* of an object depends on just two things: (1) how much radioactive material, and (2) its half life. No other quantity is in that equation. No other quantity is needed.

The radioactivity given by [17.06] is the radioactivity **Now**. "Now" means not necessarily at the time you are looking at the equation; it means at the time when the number of radioactive nuclei in the object is n_0.

"The object" can be any collection of material that you choose to examine. It can be a rock or a whole cloud of stuff in the atmosphere or an amount of some radioactive fission product in a fuel rod of a reactor.

Implications of equation 17.06

Equation [17.06] has two direct implications.

It says that the more of a radioactive stuff you have the more radioactive it will be. That is hardly surprising.

But eq 17.06 also says that the greater the half-life of that material, the *less* radioactive it will be.

A long half-life means that it will last longer, but will be less intensely radioactive in the meantime.

One way to think about this is that all n_0 of these radioactive nuclei are destined to dissociate, some time. If their timetable gives them longer to accomplish that, they need not be in a hurry, and can therefore dissociate at a lower rate. Radioactivity, recall, is a dissociation *rate*. The radioactivity, a, as it comes out of Eq 17.06 is in *dissociations per second.*

Often the amount of "radioactivity" in waste materials that need to be stored, or the amount released in a nuclear mishap, will be given in *Curies*. Radioactivity tells how "bad" it is now. But bad also involves how long it will be this "bad."

Table 17.1 A short table of radioactive isotopes

At #	*Element*	*Symb*	*Mass number*	*Isotope mass*	*decay particle*	*half life*
6	Carbon	C	14	14.0032	β	5700 yr
15	Phosphorus	P	32	31.9739	β	14.3 days
27	Cobalt	Co	60	59.9338	β,γ	5.27 yr
19	Potassium	K	40	39.9640	β	1.28×10^{9}yr
38	Strontium	Sr	90	89.9077	β	28.8 yr
53	Iodine	I	125	124.90	γ	56 days
53	Iodine	I	131	130.9061	β,γ	8.04 days
54	Xenon	Xe	125	124.90	β^{+}	17 hours
55	Cesium	Cs	137	136.9	β	20 yr
56	Barium	Ba	139	138.9	β	82 min
56	Barium	Ba	144	143.9227	β	11.9 sec
86	Radon	Rn	222	222.0176	α	3.82 days
88	Radium	Ra	226	226.0254	α,γ	1,600 yrs
92	Uranium	U	238	238.0508	α,γ	4.47×10^{9}yrs
94	Plutonium	Pu	239	239.0522	α,γ	24,100 yrs

EXAMPLE

Suppose that among the radioactive waste that must be safely disposed of, there is 1,000,000 *Curies* of radioactive barium, $_{56}Ba^{139}$, as well as 1,000,000 *Curies* of radioactive cesium, $_{55}Cs^{137}$.

Problem: Compare 1,000,000 *Curies* of $_{56}Ba^{139}$ with 1,000,000 *Curies* of $_{55}Cs^{137}$ in terms of the hazards and difficulties of finding safe storage for each.

The radioactivities of both these materials *at the present time* is given. They are the same.

The half life of $_{56}Ba^{139}$ is 82 minutes. Every 82 minutes the amount halves. In 82 minutes, its radioactivity is reduced to 500,000 *Ci*; in 164 minutes it is down to 250,000 *Ci*. In one day, it will have halved more than 17 times, and its radioactivity will be down to 5 *Ci*. After 2 days, it will be down to 30 μ*Ci*. After 3 days it will be down to about one dissociation per second. It will be harmless.

The half life of $_{55}Cs^{137}$ is 20 years. Every 20 years the amount halves. In 20 years, its radioactivity is reduced to 500,000 *Ci*; in 40 years it is down to 250,000 *Ci*. It will be 340 years before its radioactivity is down to 5 *Ci*. After 1000 years, it will be down to about one dissociation per second. It will be harmless.

If you can find a place to keep the Barium away from people for 3 days, you don't have to worry about that any farther. On the other hand, to safekeep the Cesium, you will have to find a container that won't rust, and a place where living things will not be affected by it, for many generations.

The question, of course, is this: if they both have the same starting radioactivity, what is it about the cesium that makes it so much more troublesome?

The answer is that there is a lot more of it. Not more *Curies*, but more radioactive atoms to begin with. Cesium is on a long schedule, which means that any given Cesium atom is far less likely in the next 5 minutes to dissociate. It makes sense that the sample of Cesium will have to contain many more atoms than the sample of Barium in order to have the same dissociation rate, *at the present time.*

The algebraically-minded can re-arrange Eq 17.06 so that it tells how many radioactive atoms we need to have in a sample of a specified radioactivity. In other words, find n_0 for a given a.

$$n_0 = a \times (1.44) \times t_{1/2} \qquad [17.08]$$

This makes it clear that, since the half-life of the cesium is 130,000 times that of the barium, it will require 130,000 times as much cesium. A calculation will show that the samples you are dealing with consist of 0.06 grams of barium and 7.9 kg of cesium.

Each of these amounts has a radioactivity of 1,000,000 *Ci* at the outset.

Note: The isotopes of this example are those that may occur as waste in a spent reactor fuel rod. *Ordinary barium and ordinary cesium are not radioactive.*

Look at it this way

One way to imagine how these quantities are related is in terms of a flashlight. One flashlight that accommodates a large battery, one that accommodates a smaller one. Suppose first that the bulbs are the same in the two batteries.

The large flashlight battery contains more electric charge and the small battery contains less electric charge.

If they are both 3 Volt batteries, and hooked up to the same flashlight bulb, they will burn equally brightly at first. The small and the large batteries will be having to deliver the same amount of current to light the bulbs equally. The brightness of the flashlights corresponds to the radioactivity.

We know what will happen. The small battery will not last as long as the large battery. The half-life of the batteries is not the same.

An example from real life

In the next chapter we will find out that we are all exposed to a small level of "background" radiation that can not be avoided. One contribution to this background radiation comes from the small concentration of radioactive carbon, ${}_6C^{14}$, in the atmosphere. This isotope dissociates with a half-life of 5730 years, by emitting a β particle.

Carbon-14 is produced in the upper atmosphere by the bombardment of Nitrogen in the air with high energy particles from the sun. With time, the Carbon-14 made in this way dissociates. The more there is, the more is lost through dissociation per second, while the rate of its production remains constant. So an equilibrium develops in which the rate at which it is produced is just great enough to replenish the supply. The equilibrium concentration of Carbon-14 in the atmosphere is very small, about one part in 10^{12}. That means 1/1,000,000,000,000 of the carbon atoms in the air we breathe are Carbon-14.

The Carbon-14 in the air is not very harmful to us. The more dangerous beta particles are those that are shot at us from within. After all, we are made of proteins, fats, sugars, all of which contain carbon. Approximately 18.5% of our body mass is carbon.

This carbon comes from the food we eat, which comes from vegetables and fruit, or from meat of animals who got their carbon by eating vegetables and fruit. In short, all our carbon comes from plants. Plants make protein and sugar and starch by a process called photosynthesis, in which sunlight provides the energy for the plant to make these life molecules out of carbon dioxide and water and minerals from the soil.

Problem: How many times per second do we receive an internal shot of a beta particle as a result of the carbon-14 in our bodies?

This is a *radioactivity* question. The radioactivity of the carbon-14 in our bodies *is* the number of dissociations of carbon-14 per second. Its radioactivity depends on n_0 and $t_{½}$.

1. Find n_0.

A person weighing 110 pounds has a mass of 50 kg. 18.5% of 50 kg is 9.25 kg, or 9,250 grams. How many carbon atoms is that? The atomic mass of carbon (most of our body carbon is ordinary carbon) is 12 grams. 12g is the mass of one mole, or 6×10^{23} atoms.

$$\text{no. of carbon atoms} = 9{,}250\,\text{grams} \times (6\times10^{23}/12\,\text{grams}) = 4.625\times10^{26}$$

One in 10^{12} of these is carbon-14.
$n_0 = 4.625\times10^{26} / 1\times10^{12}$, or 4.625×10^{14}

2. Find the radioactivity, "a"

The radioactivity of that much carbon-14 is found using,

$$a = n_0 / (1.44\times t_{½}) \quad [17.06]$$

We found earlier that 5730 years is 1.807×10^{11} seconds.

$$a = 4.625\times10^{14} / (1.44\times1.807\times10^{11}\ \text{seconds}) = 1780\ \text{dissociations per second.}$$

That seems like a lot. 1780 times each second a high energy electron (beta particle) is let loose somewhere in our bodies.

The fact is that the normal odds are strongly against any one of these beta particles scoring a damaging hit in our living tissue. By far the vast majority of all the radioactive particles that shoot at us miss the target, or if they hit, they do inconsequential damage. In chapter 18 we will take up the matter of how harm comes from the very small fraction of them that hit an important molecule in a living body, and the kind of harm that they do.

Most of us live long lives in spite of the background radiation, of which the carbon-14 in our bodies is a part. This does not mean that these 1780 shots per second are totally harmless. Of those people who die of cancer each year, some small number would not have gotten that cancer had they not been exposed to carbon-14. When we say someone dies of "natural causes," we include the unavoidable exposure to radiation.

There is no way to "purify" the carbon we consume. There is no way to grow "carbon-14-free potatoes."

==

Carbon-14 dating

The carbon-14 that occurs in our atmosphere has, according to lots of evidence, been there in that same concentration of $1 / 10^{12}$ for many thousands of years. This has given archeologists an excellent tool for determining the age of animal (and human) remains as well as the age of artifacts such as pottery and tools that they have dug up.

Example

Imagine a village of our ancestors 5730 years ago, in which there were artists who decorated pottery with paints whose colors came from crushed flowers. The paints were glazed, and the pots survived to be found by diggers who are our contemporaries. Several art museums today have laboratory facilities that can analyze extremely small samples for radiation.

A small chip with glazed dye is ground up, and analyzed for carbon. A sample as small as ½ gram is found to contain 50 milli-grams of carbon. This sample is placed in a radiation counter, and it is found that there is, on the average, one "count" (representing one dissociation) every 208 seconds.

If you go through the calculation of the previous example, you will find that there should be twice as many counts, something like one count every 104 seconds. What you have discovered is that there appears to be only half as much carbon-14 in this sample as there "should" be.

But, of course, there should not be as much carbon 14 as your calculation gave you, because you used a ratio of carbon-14 that is maintained by the continual replenishment of our atmospheric carbon by the reaction of solar particles with nitrogen high up there.

The carbon in the flowers of which the paints were made (5730 years ago) had that equilibrium ratio of carbon-14. But as soon as that carbon was incorporated into the pottery, it ceased to be in equilibrium with the carbon in the air, and any carbon-14 that dissociated was not replenished. In 5730 years half of that carbon-14 was lost. This is the meaning of half life.

What you have discovered is that the pottery analyzed in that museum is 5730 years old. This is called carbon-14 dating.

If the number of counts had been down to ¼ the original value, it would have meant that the carbon-14 would have halved twice, and the age of the pottery would have been established at 11,460 years. This method is good to about 4 or 5 half-lives, so the method is useful for finds aged up to about 25,000 years. After that, the amount of carbon-14 left is so small that it is hard to be sure the radioactivity is due to carbon alone.

Uranium-238, which is radioactive with a half-life of 4.47 billion years, can be used to date fossils nearly as old as the earth. The earth was formed from star stuff about 4.5 billion years ago.

The mystery of naturally occurring radium

There is yet another surprise. Where does radium come from? Radium has a half life of 1622 years. Should it not be all gone by now?

Any radioactive isotope that occurs naturally has to be either continually made, as is carbon-14, or it has be left over from the beginning of the earth, as is Uranium. The half-life of $_{92}U^{238}$ is about the same as the age of the earth, and so, about half of the uranium that came from the exploding star of which the earth is made, is still around; the other half has dissociated.

In pitchblende, the common uranium ore, there is also, however, always some radium, the first radioactive element to be isolated and identified. It was through radium that Marie and Pierre Curie discovered radioactivity.

The half-life of radium, ${}_{88}Ra^{226}$, 1622 years, is an interval of time that has elapsed far too many times since the beginning of the earth for there to be any radium left, even had the earth started as 100% pure radium.

So, it must be continuing to be produced in those dark and submerged sources of uranium ore. But how?

The fact that solves the radium mystery is that in most cases, when a radioactive isotope dissociates, it becomes another radioactive isotope, and when that dissociates, it again becomes another radioactive isotope. It is the exceptional radioactive material that dissociates into a stable isotope on the first dissociation.

$${}_{92}U^{238} \xrightarrow[\alpha]{4.5\text{ BY}} {}_{90}Th^{234} \xrightarrow[\beta]{24\text{ d}} {}_{91}Pa^{234} \xrightarrow[\beta]{6.8\text{ m}} {}_{92}U^{234} \xrightarrow[\alpha]{.25\text{ MY}} {}_{90}Th^{230} \xrightarrow[\alpha]{80\text{ KY}} {}_{88}Ra^{226}$$

$${}_{88}Ra^{226} \xrightarrow[\alpha]{1622\text{ Y}} {}_{86}Rn^{222} \xrightarrow[\alpha]{3.8\text{ d}} {}_{84}Po^{218} \xrightarrow[\alpha]{3.1\text{ m}} {}_{82}Pb^{214} \xrightarrow[\beta]{27\text{ m}} {}_{83}Bi^{214} \xrightarrow[\beta]{20\text{ m}} {}_{84}Po^{214} \xrightarrow[\alpha]{0.2\text{ ms}} {}_{82}Pb^{210} \xrightarrow[\beta]{22\text{ Y}} {}_{83}Bi^{210}$$

$${}_{83}Bi^{210} \xrightarrow[\alpha]{3\text{ MY}} {}_{81}T\ell^{206} \xrightarrow[\gamma]{4\text{ m}} {}_{81}T\ell^{206} \xrightarrow[\beta]{4\text{ m}} {}_{82}Pb^{206}$$ (stable isotope of lead)

U	Uranium
Th	Thorium
Pa	Protactinium
Ra	Radium
Rn	Radon
Po	Polonium
Bi	Bismuth
Pb	Lead
Tℓ	Thallium

Fig 17.3 Uranium Dissociation Sequence

In Fig 17.3 is shown the sequence of dissociations that Uranium-238 undergoes that keeps us supplied not only with constantly fresh radium, but a score of other short- and long-lived radioactive intermediates.

The abbreviations used in the half-life times are: BY Billion years, MY Millions of years, KY Thousands of years, Y years, d days, m minutes, s seconds, ms milliseconds. In terms of time, few of the subsequent dissociations wait more than a few days to take place, although four of the steps take years or even millions of years. The long wait is for the original U^{238}. The isotopes in the sequence that have the longest half-lives are the ones that accumulate in the greatest amounts. Those with very long half-lives are not very strongly radioactive when isolated. Those that dissociate in minutes or days may not wait around while scientists isolate and identify them. It is not a coincidence, then, that radium was the first radioactive isotope to be isolated and purified. Its half life is long enough to allow it to accumulate, and short enough to yield a strongly radioactive substance.

Highlighted in Fig 17.3 is a sub-sequence with a total half-life span of about 22 years, that takes the dissociation product of Radium, which is the gas, Radon, to the radioactive isotope of Bismuth, ${}_{83}Bi^{210}$, which then stays around for some millions of years.

At the end of the sequence is the stable isotope of lead, ${}_{82}Pb^{206}$.

How to tell what a dissociation leaves behind

When a radioactive dissociation occurs, most commonly either an alpha, beta, or gamma particle is emitted. That particle leaves, but what is left behind? Because the number of protons and neutrons has changed, the element and the particular isotope of that element may not be what it was before it dissociated. Here are the "left-behinds" of dissociation:

alpha: The alpha particle carries away two protons and two neutrons from the nucleus. The number of protons, (subscript) decreases by 2. The number of nucleons (neutrons and protons, total – superscript) decreases by 4.

Example: In the uranium dissociation sequence of Fig 17.3, the first dissociation of ${}_{92}U^{238}$ is by α emission. The nucleus that is produced will be ${}_{(92\text{-}2)}X^{(238\text{-}4)}$ or ${}_{90}X^{234}$. A look at the periodic table show that X is the element with 90 protons, and that is Thorium. The isotope produced by the dissociation of U^{238} is ${}_{90}Th^{234}$, which is itself radioactive.

beta: The negative beta is an electron. It comes from the dissociation of a neutron into a proton and the beta particle. The result is an increase of one in the number of protons (subscript) and no change in the number of nucleons.

Example: The thorium produced by uranium-238 dissociation, ${}_{90}Th^{234}$, dissociates by β^- emission to the isotope ${}_{(90+1)}X^{234}$, which turns out to be Protactinium, ${}_{91}Pa^{234}$.

The positive beta is a positron; this dissociation occurs rarely and is often accompanied by capture of an orbital electron.

gamma: The gamma ray is a photon, and releases energy while leaving the nuclear makeup unchanged. γ emission often accompanies other kinds of emission.

Radon

Many homes have been found in the last 25 years to harbor dangerous levels of the chemically inert gas, Radon.

Although most homes do not have the potential of a uranium mine under them, uranium in very small concentrations is widespread. This includes the clay or rock upon which many residences are built. Where there is uranium, there is also the whole collection of dissociation products shown in Fig 17.3, including radium and radon.

Radon is the only one in the whole sequence that is a gas. This means that the purification of radon is done automatically, through the gathering of radon bubbles. Also, once the Radon gas has collected, it makes its way upward through the moisture in the ground, and through cracks and fissures. The part of the gas that bubbles out to the open air is not of concern; it is diluted by the atmosphere.

The part that enters the home through cracks and pores in the foundation, collects in the basement air, and tends to accumulate there, because it is 7 times denser than air. As air currents stir it up, it spreads to the rest of the house, but is normally in no hurry to leave, especially in the winter when the heating system may circulate the same air many times.

The concentration of Radon is measured in pico-Curies per liter of air. The prefix, pico, means 10^{-12}, or trillionths. In the United States, the recommended maximum tolerable year-round average level is 4 p*Ci* / liter. The general average found in American homes is about 1.5 *pCi / liter.*

Radon in homes was first noticed in an area with relatively high uranium concentration in the ground. Some homes were found to have concentrations 100 times the recommended maximum.

The good news is that it is usually easy and inexpensive to reduce the Radon level in a home that has too much. The sealing of cracks and spaces around pipes in the foundation is often sufficient. If that doesn't suffice, a small fan that creates an air pressure gradient in the lowest areas of the house, will draw the Radon-laden air to the outside, and replace it with fresh air.

Small kits for measuring Radon level are available at most health departments or at hardware stores. The kit contains an absorbent material that traps the radon dissociation products, and the whole package is sent in for analysis. Because Radon is a seasonal event, it is often recommended that a kit be hung in the basement for an entire year.

The fact that the dissociation products of Radon are solids is a disadvantage. Each Radon atom that disintegrates is responsible for four further dissociations within an hour or so, leaving ${}_{82}Pb^{210}$, an isotope of lead that can get attached to the lining of the lung and stay there radiating for 22 years.

While the idea of inhaling this Radon laden air sounds frightening, it is alarming only if large concentrations are found. If the average lung holds on the average one liter of air, at the tolerable maximum level, the lung will be exposed to 4×10^{-12} *Ci* of radiation, which is $(4\times10^{-12})\times(3.7\times10^{10})$ or about one dissociation every 7 seconds. This is a much smaller number than the number of unavoidable dissociations from carbon-14, but may be equally damaging because half of the dissociations from the radon sequence are alpha particles, which are more damaging to living tissue than the beta particles given off by carbon-14.

Problems

17.1 Using information in Table 17.1 and in a periodic table, tell what isotope is produced by the radioactive dissociation of the following: (A) Phosphorus-32; (B) Radon-222; (C) Uranium 238

17.2 Calculate the radioactivity of 1 kg of Uranium-238.

17.3 Find the radioactivity of 1 milli–mole of Cobalt^{60}. This is the material most commonly used in radio-therapy for cancer. Now find the radioactivity of the same number of atoms (1 milli–mole) of radium. Radium is reputed to be so very powerfully radioactive. Why are the radium atoms less radioactive than the cobalt^{60} atoms?

17.4 Show that the radioactivity of 1 gram of radium is indeed equal to 1 *Ci*, as it should be because the definition of the unit is that one Curie is the radioactivity of 1 gram of radium.

17.5 (A) Express in Curies the (radio-)activity of 1 gram of Iodine-131.
(B) Express in Curies the (radio-)activity of 1 gram of Strontium-90.
(C) Explain why these two radioactive fragments from uranium fission are particularly hazardous if ingested by humans.

17.6 The atmospheric concentration of Carbon-14 (in the carbon dioxide of the air) is one part in 10^{12}. If the carbon in a skeleton you dug up had in it 1/16 part in 10^{12} of carbon-14, about how long ago was this animal alive and breathing?

17.7 A small amount of potassium (K) is necessary for proper muscle function in your body. Without it, you get muscle cramps. 0.35% of your body mass is potassium. 0.00118% of naturally occurring potassium is K^{40}. Find how many dissociations per second (Bq) you get inside your body from this radioactive potassium. Use 50 kg as your body mass. These dissociations are in *addition* to those of C^{14}. Which contributes more dissociations, C^{14} or K^{40}?
What is the total radioactivity due to both C^{14} and K^{40} in your body?

17.8 The Report of the President's Commission on the Three Mile Island accident concluded that between 2.4 million and 13 million curies of radioactive Xenon and Krypton were released into the environment as rare gases. (A) What are some things you would want to know in order to evaluate how serious this is?

Suppose you found out that 2.0 million Curies of Xenon-125 were released. Xe^{125} dissociates, producing Iodine-125, which, is also radioactive, emitting a gamma.

(B) How many *grams* of Xenon were released? [Hint: Use *the* equation]

(C) How many *grams* of Iodine-125 are ultimately produced? (Don't make this hard)

(D) How many *Curies* of Iodine-125 are produced?

(E) Compare the "danger" to humans of this many Curies of Iodine-125 with the danger posed by the original 2,000,000 Curies of Xenon-125. The answer to this is not necessarily simple or easy. It may involve several factors.

Ans:

1. (A) ${}_{15}P^{32} \rightarrow \beta + {}_{16}\mathbf{S}^{32}$ (B) ${}_{86}Rn^{222} \rightarrow \alpha + {}_{84}\mathbf{Po}^{218}$ (C) ${}_{92}U^{238} \rightarrow \alpha + \gamma + {}_{90}\mathbf{Th}^{234}$

2. $\mathbf{1.24 \times 10^7}$ **disso/sec** = $\mathbf{3.35 \times 10^{-4}}$ **Ci** or **0.335 mCi**

3. Co^{60}: 2.51×10^{12} Bq= **67.8 Ci** ; Ra^{226}: 8.26×10^{9} Bq= **0.223 Ci**

5. (A) 1 g Iodine-125: 6.89×10^{14}diss/sec = **18,600 Ci**

(B) 1 g Strontium-90: $5.87 x 10^{12}$diss/sec = **159 Ci**

(C) When ingested, strontium is incorporated into muscle and bone in place of calcium, where it remains for years; iodine is incorporated into the thyroid

6. 1/16 = (½)(½)(½)(½) or $(½)^4$. 4 half-lives or 4 x 5,700 or **22,800 years**.

7. **546 disso/sec from $\mathbf{K^{40}}$**. Total from C^{14} and K^{40} **2284 disso/sec.**

8. (B) $n_o = (a)(1.44 \times t_{½}) = 6.52 \times 10^{21}$ atoms = **1.36g of $\mathbf{Xe^{125}}$**

(C) **1.36g of $\mathbf{I^{125}}$**

(D) The *activity* of this iodine 9.37×10^{14} diss/sec= **25,300 Ci**

18. Radioactivity and Exposure

The analogy of radioactivity with the brightness of a flashlight has a further message.

When our interest focuses on how many dissociations occur per second in an object, then we are talking about its *radioactivity*. Radioactivity corresponds to the *brightness* of the flashlight.

We may also want to focus not so much on the brightness of the flashlight, as on the level of illumination that it creates on an object that it shines on. If it shines on a page of a book that is few inches from the flashlight, the chances are good that the text on the page will be readable. If the book is across the room, the page may be much more dimly lit, even though the flashlight is the same.

In much the same way, there are many factors that determine how "brightly lit" an object is with radioactivity. How "brightly lit" something is, whether with light or with radioactive particles, is a measure called, "exposure."

In the case of radioactivity, the term "exposure" is used in two ways that need to be distinguished. One is the amount of radiation that is received by an object per second. This is, strictly speaking, called the "rate of exposure." The word "exposure" itself is defined in terms of a cumulative effect.

This distinction is significant especially if the object receiving the radiation is human and we are concerned about the damage done. Obviously, since the radiation damages various parts of the body in proportion to the total number of alpha, beta, or gamma "hits" that are suffered, there is more "exposure," and more damage, if the person is exposed for a longer time.

The badges that are worn by employees in hospitals and other facilities where they are exposed to radiation contain a piece of film.

Some small fraction of the radioactive particles that pass through the film ionize granules of colorless silver bromide in the film, producing gray colored metallic silver, which shows up as a darkening of the film when it is developed. The longer the badge is exposed to radiation, the darker will the film be. The badge exposure is proportional to the cumulative hits that the person wearing it is exposed to. At the end of the month, the film is removed from the badge, and the coloration of the developed film is a record of the exposure the wearer has had during that month.

A note about "radiation"

There is sometimes misunderstanding about the term, *radiation*, because it has different meanings. There is a great deal of electromagnetic radiation that, in moderate quantity is almost totally harmless. This includes the visible light from the sun, heat radiation, and radiation from low frequency radio, which are in a different category from the radiation referred to in connection with radioactivity.

The radiation exposure from the various radioactive dissociations, and from high energy electromagnetic radiation such as the ultraviolet rays from the sun and X-rays, are different because they have the ability to *ionize* atoms and molecules, that is, to strip electrons from their orbits. Thus they can change the chemical makeup of single molecules with a single hit from a single alpha, beta, or gamma. If the damaged molecules are DNA molecules in the cells of a living organism, the damage can lead to cancer and alteration of the gene makeup of offspring.

When we speak of radiation in this chapter, we speak only of *ionizing* radiation.

"Counting" with a Geiger Tube

There is not a simple equation by which the amount of exposure to radioactivity that a person has experienced can be calculated. Exposure is generally something that is measured.

The device used to measure exposure is a Radiation Counter called the Geiger-Mueller Counter, or, just "Geiger Counter." Like the photographic badge, the radiation counter tube is placed where the exposure is to be measured. The tube itself can be small, an inch in diameter and about ten inches long. It is where the radiation particles are counted.

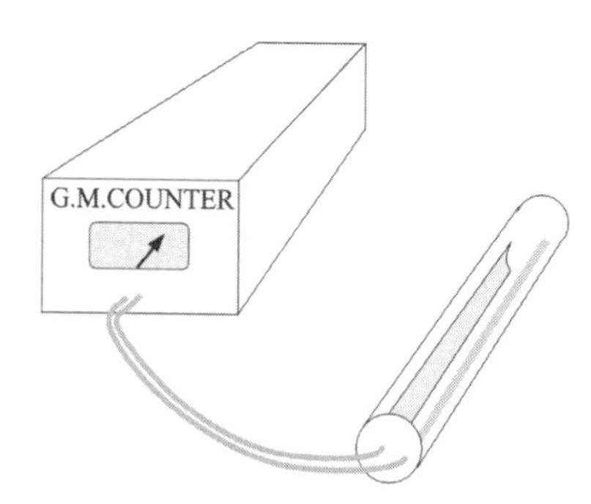

Fig 18.1 Geiger-Mueller tube and counter

What is counted is the ionization of a gas, usually at low pressure, inside the tube. If it is not ionizing radiation, it doesn't get counted here. There is a high voltage between two metallic electrodes inside the tube. There is normally no current, because the gas is a non-conductor, except when it is ionized. When an alpha particle, for example, passes through the tube, it ionizes one or more of the atoms of the gas, creating (temporarily) some ions that can conduct electricity. There is a short burst of current, which passes through a cable to an electronic circuit box that may be some distance away, where that blip of current, when amplified, causes a small click of sound. When you hear the "counts" in the Geiger counter, you are not hearing the radiation; you are hearing the sound that is triggered by the brief bursts of current in the tube.

If a Geiger counter is placed in an area where there is no additional source of radiation, it records just the background radiation. With a small laboratory counter, one hears clicks irregularly on the average of one every few seconds. If a source of radiation is nearby, the clicks will be more frequent, and can become just a continuous noise, in which it is no longer possible to count clicks per second.

The electronics usually is connected also to a circuit that adds counts and produces a voltage proportional to the number of counts per second. This activates a meter, where a needle or a digital output indicates an average number of counts per second. There are usually settings for several ranges of values that will allow the meter to read without going off scale.

REMs and their origin

There are wide variations in how radiation counters are constructed and what they measure. Obviously the bigger the tube, the more particles will cross it, and the more clicks will be recorded. But also, counters may be designed specifically to count only gamma radiation. Such counters will use a tube shielded with a thickness of metal that will stop alpha and beta particles, but will pass gamma rays. If a counter reports 400 counts per second, that in itself does not tell the whole story.

Counters do not measure *Exposure*. They measure *Rate of exposure*.

Exposure is cumulative

Exposure is *not* the number of ionizations *per second.* Exposure is a *cumulative count of ionization*. The longer you are exposed, the greater is the exposure. A Geiger counter read-out does not reflect exposure; it tells the *rate of exposure*. If you are exposed to radiation, the damage is done by the total exposure, not the rate of exposure.

The *cumulative* nature of *Exposure* is in contrast to the concept of *Radioactivity*, which is a *rate* of dissociation.

The practical unit of radioactive (cumulative) exposure is called the **REM,** which stands for "Roentgen Equivalent Man." That requires some explanation.

The original, objectively defined unit of radioactive exposure is the *Roentgen*, named for the discoverer of X-rays. It is defined in terms of the number of ionizations produced in one cubic centimeter of dry air.

There are numerous drawbacks to the *Roentgen* as a unit of exposure. First, the object being exposed is usually not air, and its response to radiation may differ from that of air. Second, air, and other materials, differ in their ionization susceptibility to alpha, beta, and gamma, so that an exposure of 2 *Roentgen* of alpha may translate to targets other than air differently from exposure of 2 *Roentgen* of gamma. There are other problems as well.

There is a history of attempts to adapt the measurement unit to various applications. The RAD, and the GRAY, which is 0.01 RAD, are units that readers will encounter in the literature, but which are not necessary for this discussion.

Eventually a unit was developed for use when the main interest is in the damage to living human tissue (hence the "equivalent man" part of the name.[1]) This unit was adjusted for ionization *in the human body*, penetration through the *human skin* and into the *vital organs*. This unit was equalized so that one REM of exposure, whether it be alpha, beta or gamma, whether it be internal (in the lungs) or external (through the skin), would be likely to produce a uniform degree of damage.

Such a unit can not be exact. Different tissues – lung, liver, skin, bone, eye – will not suffer from exposure equally. Age, size, radiation tolerance, and other individual differences among exposed

---

[1] This nomenclature was chosen before it was acknowledged that using "man" to mean "man or woman" is sexist. (It has not been changed to "Roentgen Equivalent Person.".)

people also affect the damage to be expected from exposure. The immediate effects do not necessarily track the long term effects.

But, overall, the REM has come to stay as the unit for measuring radiation hazard for the human species.

How bad is a REM? How serious is it to be exposed to one REM?

This question has to be answered in two parts, because there are two distinctly different ways in which radiation affects living things. They are referred to as *immediate* and *delayed* effects.

1. Immediate Effects: Burns, inside and out

Massive energy damage is immediate, and not unlike the damage from non-ionizing but penetrating irradiation such as occurs in a microwave oven.

Except for scarring, internal and external, recovery from moderate immediate effects can be complete, and no long term consequences are expected.

Immediate effects consist of destruction of living cells by the sheer impact of energy. When so much of a cell is destroyed or incapacitated that it can no longer reproduce or function, that cell has been burned. Because gamma and, to some extent beta, radiation can penetrate, it can burn not only the skin, but also the inner organs, liver, heart, muscle, nerve.

The body can tolerate a certain amount of this burning. If the cells disabled by burning do not shut down the functions of the tissue burned, the remaining cells will eventually multiply to replace the ones lost, and there will be reasonable or even complete recovery.

The effects of cell destruction of this kind are generally treated with the usual medical tools: covering damaged areas, skin grafts, rest, protection against infection. More sophisticated methods include bone marrow transplants to regenerate the ability to make red blood cells.

Table 18.1 Immediate effects of radiation exposure*

above 8000 REM	*fatal immediately or within a day*
650 to 8000 REM	*delayed fatality, usually within several weeks*
150 to 650 REM	*"radiation sickness", with recovery*
100 to 150 REM	*mild radiation sickness, bone growth retardation*
less than 100 REM	*temporary impairment of immune system*

There is great variability among individuals. These are guidelines only.
* *Source: Army Field Manual*

There is one major difference between heat burns and radioactive burns. The source of heat burns is usually visible, and one has a sense that at least to a limited extent one has the capacity to avoid such exposure. The source of radioactive burns is invisible. The firefighters who went atop the now roofless reactor building at Chernobyl to hose down the burning graphite had suits to protect them from heat burn, but were not adequately protected from radioactive exposure, and were probably unaware of it until it was too late.

In extreme situations, such as warfare, there are, in addition to direct radiation burns, secondary burns from fires started by the nuclear blast.[2]

---

[2] Most local emergency preparedness agencies publish pamphlets with data on likely radii of influence of nuclear bomb blasts. An extensive discussion of biological and medical effects of radiation exposure is found in P.Craig and J.Jungerman, *Nuclear Arms Race*, McGraw-Hill, New York, 1986

2. Delayed effects

Delayed effects of radioactive exposure are those due to genetic damage. This is damage to the genetic material in an individual cell. It consists of chemical damage produced by radiation-induced ionization somewhere in the DNA chain that carries the blueprint for the cell, and for the entire organism.

The injury to a DNA chain is a small but far reaching event. Although the injury can come from a single radioactive particle that hits the DNA molecule in a single spot, the injury is propagated through all the future generations of that cell because the DNA molecule is a blueprint. If the cell has become a cancer cell, all its progeny will be cancer cells. If the cell is an egg or sperm cell, its progeny are the children born from that cell, and the tiny injury to the blueprint is capable of causing a defect in the child.

In the sense that such an injury is due to a single radioactive particle, there is no such thing as a safe level of radioactive exposure.

The statistics of the likelihood of delayed effect damage is the outcome of a precarious balance between the enormous odds *against* any single radioactive particle having a DNA encounter that causes an injury that is propagated, and the continuous barrage of radiation, both internal and external, that every human body endures, even in the absence of human-made exposure.

We have already calculated in Chap 17 that a human body is the target of thousands of emitted betas from carbon in the body, plus alphas from radon, plus numerous other sources of unavoidable background radiation.

The odds against any one radioactive particle causing an ionization in a sensitive point in a cell is due to several factors:

(1) Most encounters between a radioactive particle and the living organism occur in the 99-plus percent of a cell that is not DNA.

Damage in parts of the cell other than its DNA injures that cell only. If that cell propagates, it will produce normal offspring because the DNA was unaffected. If it dies, it will be replaced by cell division of healthy cells.

(2) If there is radiation damage to DNA, it is usually fatal to the cell's ability to reproduce. Disabling the cell's ability to reproduce entirely is a minor injury, restricted to that cell only. When that cell dies, its disability will die with it. There will be no propagated cancer, and there will be no child born of that sperm or egg cell.

(3) Many injuries to the blueprint for a cell are physiologically benign. Only very specific injuries cause the particular type of alteration in a cell that causes it to lose the ability to limit its reproduction and become a cancer cell.

It is because of these factors that, in spite of all the radiation humans are exposed to, most people survive quite well for many years.

Cancer

Every living cell contains in its cellular (not atomic) nucleus a blueprint for its own design. This blueprint is encoded in an enormously long chain of little components, just four different nucleic acid subunits. It is like a magnificent work of literature written in arrangements of the characters of a four letter alphabet.

If a cell has become cancerous through an alteration of the DNA chain, what makes it a cancer cell?

A cancer cell has lost the part of its DNA blueprint that shows it how to control its rate of reproduction. Normal cells respond to signals from the cells around them that there enough of that particular cell for the time being, and they need not be dividing until more replacements are again needed. Cancer cells are unresponsive to these signals, which are probably chemical, and they continue to divide and reproduce without restraint.

The cancer cell multiplies constantly, and soon there are many more of these than of normal cells. They outgrow the physical space allotted to them, and they become a lump, called a tumor. If a cell breaks off the lump, it can be carried by the blood or the lymph to a distant place, where that one will start the tumor process all over. This process is called metastasis. If not stopped, it will kill the person.

It is thought that for a cell to become cancerous requires not a single point of damage, but several separate alterations, very specifically located in the DNA chain. These injuries may be from chemical causes; a cancer inducing chemical is called a carcinogen. Or, they may be caused by a cosmic ray or some other source of background radiation, or by radiation from artificial sources.

Much is still not known about the details of how cells become cancerous. One puzzle has always been that older persons are more susceptible to cancer than younger ones. If cancer were caused by a single ionization event in a single cell, then this should occur with the same likelihood in a younger as an older person.

A factor that may favor the younger person is the involvement of the immune system, which characteristically diminishes in efficacy with age. It may be that many cancer cells are recognized as strangers by the immune system, so that in the younger person there is a greater likelihood that any cancer cell that is formed will be picked off by the immune process.

But the diagnosis of "pre-cancerous" cells has supported the idea that multiple injuries in very specific locations along the DNA chain are required for a cell to lose control of its reproduction. Over time, a person may accumulate a population of cells in which one or two of such injuries have occurred. Only when one of these partially damaged cells incurs the third (or fourth or fifth) required injury, presumably later in life, is the path to cancer finally completed.

This means that a particular radioactive exposure may not ordinarily "cause" cancer. What *can* be said, if this model is true, is that had it *not* been for a particular radioactive exposure, a cell that did become cancerous due to three injuries, would have remained with only two, and not become a cancer cell. The three hits may have occurred many years apart, and it can not be said which of them "caused" cancer. All three played a role, but the absence of any one of them could have *prevented* the final outcome.

Birth Defects

The blueprint for a new child is in the combination of the DNA of a sperm and an egg cell. Birth defects occur when a sperm or egg cell is damaged in such a way that the blueprint for the baby has a mistake.

Much of what has been said about the creation of a cancer cell applies also to the creation of a child with a birth defect. Most particles don't make a hit; most hits disable the genetic cell, so that no baby is produced at all; most hits that affect the characteristics of a child that is born are harmless, either non-functional, or affecting a characteristic that is non-essential, such as earlobe size.

That small fraction of hits that have a severe but not fatal impact on the genetic blueprint, give rise to birth defects. As with other effects of radiation hits, it is the combination of very low probability with an enormous number of radioactive hits that produces the birth defect.

Delayed effects are measured by likelihood

The way in which normal cells become cancerous makes it very difficult to make a table for the delayed effects, linking exposure to damage (like Table 18.1 for immediate effects). If the multiple injury model is correct, it would account for the fact that the effects of radiation in an affected area surface in the cancer statistics many years later. Radiation exposure of a large population over a short period of time could produce many first or second injury cells, which would show up as a high incidence of cancer many years later, due to third injuries *from other causes*. These would be cancers triggered by a normal incidence of other factors, such as background radiation, adding a third injury to a greater than normal population of two-injury cells. Without the radiation incidence many years earlier, most of these two-injury cells would now be zero or single injury cells, that would not become cancerous with one more hit.

The cancer story is one in which the statisticians necessarily play a large role. Many studies have been made of the survivors of Hiroshima and Nagasaki over many years. Smaller populations have been studied for incidence of certain kinds of unusual cancers that seem to be associated very strongly with radiation exposure. For example, the statistical rise in cases of a particular type of childhood leukemia in eastern upstate New York and Massachusetts has been linked to a week of heavy rainfall from a fallout-laden cloud pattern that had traveled at higher altitudes from Nevada atomic bomb tests to the east.

From a variety of such statistical studies, a relationship between statistical likelihood and exposure has been deduced. It makes sense that the more people are exposed the more cancer. It also makes sense that the greater the exposure of each of those people, the more cancer. This is expressed by saying that there is some number of *people-REMs* that will result in a statistical probability of an added cancer case occurring. The data appear to support the conclusion that for every 10,000 people-REMs of exposure, there will be 1 to 2 cases of cancer that otherwise would not have occurred.

10,000 people-REMs could be 10,000 people exposed to one REM, or, 20,000 people exposed to 0.5 REM, or 5,000 people exposed to 2 REMs, and so on. It means that one person exposed to one REM increases his or her chance of contracting a cancer that he otherwise would not have contracted by 1 to 2 in 10,000, which is a 0.01% to 0.02% additional lifetime risk.

Example 1

It is estimated that 200,000 people living in the Gävle region of Sweden received an *added* exposure of 0.4 REM over a period of a year after the winds blew rain clouds over their area from the nuclear plant accident at Chernobyl, a thousand miles away, in 1986. How many cancers can be expected in that population that otherwise would not have occurred?

200,000 people receiving an added exposure of 0.4 REM is 80,000 people-REMs. At the rate of 1-2 cancers per 10,000 people-REMs, this is 8 to 16 cancers. In a population of 200,000 the number of persons expected to contract cancer in their lifetimes is much greater than 16. The overwhelming cancer cause in that area is not the fallout from Chernobyl, and so it would be wrong for any individual to place the cause of his or her cancer on that accident. Yet there are 8 to 16 individuals there who would be right to do so; it's just not possible to say which 8 to 16 people. The increase in the cancer rate in that area is not considered a major catastrophe, yet for 8 to 16 of those residents, whoever they are, it is a great tragedy.

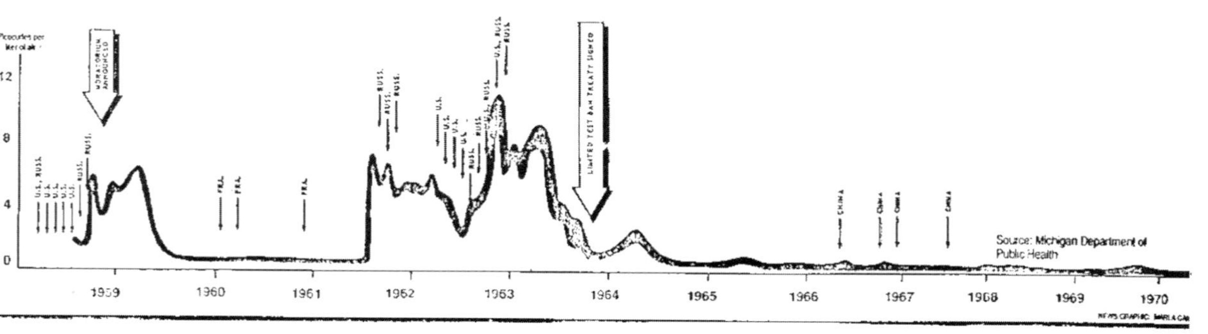

Fig 18.2 Unintended result of monitoring the Monroe (Michigan) nuclear plant for radiation leakage was this striking profile of the effect of radiation release half way around the world. Thin arrows point to bomb tests (Country conducting them is shown, barely readable – U.S., RUSS, FRANCE, CHINA). Heavy arrow at left is "Moratorium announced"; heavy arrow at right is "Limited Test Ban Treaty Signed." Before and after periods of bomb testing the graph shows radiation level in Monroe to be at background level. (Vertical axis is in picocuries per liter of air.)

The newspaper clipping of Fig 18.2 shows how far from its origin the effect of fallout from atmospheric atomic bomb testing is felt. The graph shows that the atmospheric background radioactivity in Monroe, Michigan, is about ½p*Ci* / liter of air during the years when there was little or no testing. During the years of bomb tests, all of them thousands of miles away, the background radioactivity in Monroe rose to 4 and even as high as 8 p*Ci* / liter.

The monitoring was done in Monroe to measure radioactivity from the operation of the Fermi nuclear plant nearby. What it showed was that there was very little radioactivity of local origin, but a clearly documented increase in environmental radioactivity as a result of atmospheric testing of nuclear weapons half way around the world. There is no easy way to translate this effect into REMS for the exposed population of the Northern Hemisphere.

A much more dramatic, and immediately documented effect, occurred in Scandinavia after the accident at the nuclear plant in Chernobyl, Ukraine (then in the Soviet Union) in 1986.

The reindeer of Lapland

In the northern part of Scandinavia, above the arctic circle, lies a land of 30,000 people called Lapland. Parts of it are spread over Norway, Sweden, Finland, and Russia. The same cloud that passed over Gävle rained down also on large parts of Lapland. It created no dramatic increase in the cancer rate of the Lapps, and might have gone unnoticed, were it not for the science of protecting the public from food contaminated by radioactive isotopes above a certain concentration.

The Lapp economy is dominated by the herds of reindeer that provide for the people of Lapland food, clothing, and material for their tents, as well as their main export, with which to buy manufactured goods from other countries. Reindeer graze on a plant with shallow roots called reindeer moss. The shallow roots caused the slightly radioactive rain from Chernobyl to be absorbed quickly into the reindeer moss, and from there concentrated into the reindeer themselves. The reindeer suffered no symptoms, but their meat did not pass inspection, because of its content of radioactive strontium. This is a particularly dangerous isotope for human consumption, because its chemical similarity to calcium causes the human body to incorporate it in muscle and bone, where some of it can remain for most of its 30 year half-life.

Almost the entire reindeer population of Lapland had to be destroyed and buried under signs saying, "Danger: radioactive material underground." For the proud and self-reliant Lapps, this was an unexpected invasion that has proved a tragedy for their economy and their culture.

Table 18.2 Exposure and Delayed effects

Recommended maximum permissible exposure per year	0.5 REM
Background exposure due to unavoidable sources	0.1 REM
Exposure from one chest X-ray	0.005-0.015 REM

1 to 2 cancer cases, statistically, for every 10,000 people-REM

Exposure is often expressed in milliREM.
1000 mREM = 1 REM

A question often asked is, "How many Curies causes one REM?" There is no answer to that, in part because one quantity is a rate and the other is cumulative. But also because there are the many factors that contribute to the extent of exposure.

A million flashlights will not light up your book if the book is not in their beam. Likewise, a million Curies will damage you only to the extent that you are in the path of the emitted particles.

The relation between brightness and exposure involves many factors. For a source with a given radioactivity, the exposure that an object experiences depends on the distance to the target object, the duration of exposure, any obstacles that may prevent radioactive particles from reaching the object, the kind of particle (alpha, beta, gamma), the size and nature of the target object, the persistence of the radioactivity, which may decline with half-life or by movement of winds or the flow of water in which it is dissolved, and so on.

In humans one of the main factors determining their exposure from a given quantity of radioactive material is whether and how it enters the body.

If the source of radiation remains outside the body, it is referred to as "external" contamination. This includes fixed sources that may be inches or miles away, contamination of skin, clouds, soil, rivers, vegetation.

If the radioactive material enters the body, it is referred to as "internal contamination." What effect it has depends on the manner in which it enters the body.

If it is inhaled, does it, like radon, leave dissociation products that are solids and cling to the lung wall and continue along a radioactive decay sequence? If it is ingested, how is it metabolized? Does it pass through within a day, or does the body concentrate and deposit it, as, for example iodine-131 going into the thyroid gland as if it were ordinary iodine-127, or strontium-90, which goes in for calcium in bone and muscle?

Radiation sickness and treatment of cancer

It is ironic that cancer is treated by irradiating tumors, when radiation is one of the main causes of cancer. But the cause and the treatment come from the two different kinds of biological effects of exposure to radiation.

Cancer is a delayed effect, induced by a pinpoint single event of ionization that occurs when one radioactive particle hits one DNA molecule in a particular place.

The treatment, on the other hand, involves cell destruction due to intense exposure, targeted if possible on the tumor. It is done only if the tumor is not operable, usually because it has so infiltrated healthy tissue as to make surgery inadvisable.

Radiation can work in such cases because it can differentiate between cancerous and healthy cells even if these are quite intertwined. It is able to do this because the cells that are most vulnerable to massive cell destruction are cells that are in the process of dividing. Cancer cells are constantly dividing. This makes them more vulnerable than their healthy neighbors to being destroyed by radiation.

Normally the most furiously reproducing cells in the body are the cells in the hair follicles and the cells in the lining of the digestive tract. (This is why hair continues to grow even when the body has stopped growing.) Red blood cells in the process of being produced in the bone marrow, to replenish those in the circulating blood, are also among those that are dividing avidly.

This is why the principal symptoms of radiation sickness from massive radiation exposure are intestinal upset with diarrhea, hair loss, and the loss of red cell production in bone marrow. There is recovery from moderate radiation sickness. If there are enough healthy cells that survive, the damaged cells just fall away, the digestive symptoms disappear, hair grows back, and the supply of red blood cells is restored.

Chemotherapy for cancer works the same way. A drug is administered that targets cells that are in the process of dividing. It is no surprise that the side effects of both chemotherapy and radiation treatment are the same three symptoms that accompany exposure to large doses of radiation.

Problems

18.1 The internal radioactivity due to radon in the average person living in an average home is about 1.5pCi. You are told that this causes about as much cancer as an *exposure* of .025 REM/yr, or about 1/20 of the maximum recommended allowable exposure. Calculate, based on this information, the estimated number of cancers produced in the U.S. each year by Radon. (population about 250,000,000)

18.2 It has been estimated that during the months after the explosion at the Chernobyl reactor, about 200,000 persons in the near area received exposures of between 5 and 50 REM. Assume that the average dose received by those with only long-term effects was 20 REM, estimate the number of cases of cancer that will ultimately result from the nuclear disaster.

18.3 In a city of 500,000 people, each person receives annual chest X-rays from age 20 to 50. How many cancer cases would be expected as a result of these X-rays in that population? Make some comments about the risk that is incurred in comparison with the diagnostic value of the X-rays.

Ans:

1. *given* average exposure due to Radon in the U.S. is .025 REM/yr.

(.025 REM/yr) × (250,000,000 people) = 6,250,000 people-REM/year

625–1250 cancers/year

2. **400 to 800 cancer cases**

3. Over the range of X-ray exposures (Table 18.2), **7 to 45 cancer cases**

PART V

Risk Assessment

19. Getting the Information

On January 13, 2000, The New York Times ran a front page story under a gripping color photo showing grim-faced Japanese citizens of all ages, kneeling, led by a young woman in a protest related to four nuclear plants in their village of Takahama. The headline atop the story was, "Accident Makes Japan Re-examine A-Plants."

The story includes this paragraph: "The brief meeting between the industry chief and the governor illustrates how sharply the ground has begun to shift under Japan's electric utilities since workers set off the accidental chain reaction at Tokaimura, 70 miles north of Tokyo."

Japan's worst nuclear accident (in September 1999) had killed one worker and exposed scores of people to radiation.

In February, 1998, the Associated Press distributed a story which began, "A study of radon blames the naturally radioactive gas for 21,800 American lung cancer deaths a year." This was not a report from a fringe organization; it came from a National Research Council study committee headed by Dr. Jonathan Samet, a Johns Hopkins University professor.

The information upon which the average world citizen forms a basis for assessing the risk of radiation exposure is made up of these periodic news items, plus an awareness of the background against which the news events occur.

There is the knowledge that there are still thousands of missiles in the U.S., Russia, and several other nations, launch-ready, carrying armed thermonuclear warheads with a destructive power equivalent to that of over a million tons of TNT (dwarfing the 12,000 tons of TNT equivalent of the bomb that leveled Hiroshima).

There is the memory of high profile nuclear accidents, such as the near-meltdown in 1979 at Three Mile Island, near Harrisburg, Pennsylvania, in the eastern U.S., and the explosion in 1986 of one of the reactors at Chernobyl, near Kiev, Ukraine in the European part of what was then the Soviet Union.

There are then the dozens of times that the public has learned of negligence in the handling of some of the tons of artificially produced radioactive materials. They know that this negligence has caused injury, disease, and death. Just one such incident is the case in which a Texas hospital disposed of a used cancer treatment radiation machine, without first removing the partially used pellets of radioactive Cobalt. The salvage industry sold and resold parts of this machine, and the pellets ended up rolling down the sidewalks of Aldama Street in a little Mexican town, where children played with them.[1]

There is awareness of the mysterious movement of this danger, that can slip unseen, unheard, with no odor and no taste, far and wide.

Negligent exposure to radioactivity in "downwind" farms and homes near the locales of atomic bomb tests and nuclear fuel production facilities makes news now and then as class action lawsuits pass through the courts.

In the aftermath of the Chernobyl accident, there were daily maps of the spread of the high altitude air currents that had become the carriers of the 300 million-plus Curies of radioactive fission product from the core of the exploded reactor. The cloud spread northwestward, and was spotted in Sweden, the first that anyone outside the Soviet Union learned of what had occurred. But over days and weeks, the cloud circled around, back over much of European Russia, then south to Greece and Italy, then back around

---

[1] "Hot" ("The incident on Aldama Street that became North America's worst radioactive accident") by Susan West, *Science 84*, December 1984 .

to central Europe. In all some 200,000,000 people received significant exposure, with a probable one to ten thousand cancer deaths ultimately that would not have occurred in the absence of that accident. Who those individuals are, out of all those who died of cancer, will never be known.

Italy banned the sale of fish from the lakes and rivers of northern Italy after rain brought to that beautiful mountain resort area the fallout from Chernobyl, over a thousand miles away. The people of Italy have since voted to prohibit nuclear power plants in their country.

Documentation of fallout from well-known events make news, and then fade into background.

Occasionally the reality that radioactive fallout knows no boundaries is brought home in data gathered routinely, as in Fig 18.2. Such data are often more alarming than the news from the major calamities, because for many the major events are "somewhere else," and eventually fade into the past.

The mission of this book

The mantle of "matter-to-energy conversion," and its tie to Einstein and relativity, has done a thorough job of discouraging the general public from trying to understand nuclear energy.

We hope we have lifted that curtain and given non-expert readers a basis for making judgments based on facts rather than on someone else's interpretation of the facts, and to bring informed judgment to the issues. This will happen imperfectly, to be sure, but perhaps on a par with the level at which ordinary people make judgments on economics and other public policy issues which are important to them.

The issues of nuclear energy are after all not matters of how to design a better HPI pump, but on what to do about the dangers of mistakes in the launching of weapons, what sort of public interest regulation there is to be over nuclear operations, how to store radioactive waste, above all what risks to take.

It is not the mission of this book to evaluate the literature that is born primarily of bias. This could be the subject of another whole book. It is perhaps realistic to think that with the knowledge gained from this book, the reader can him/herself better sort out the more reliable from the less reliable in the many books and other articles of persuasion that they will encounter.

For most this search will be a voluntary activity. Others are immersed in the issues through no choice of their own. Science teachers are among those. They are targeted relentlessly by one additional source that is perhaps worthy of special mention. The American Nuclear Society, a legitimate industry and professional organization, sends unsolicited magazines, catalogs, audio-visuals, and even graduate students, to science teachers, with the aim of reaching children through their teachers, from kindergarten through high school, with the message of the benign atom. They encourage, for example, giving ANS materials as awards to student "scientists of the day." One of these suggested awards, which teachers can order at a nominal cost in large quantities, is a

Fig 19.1 Badge available to teachers.

three-inch sticker (Fig 19.1) that can be worn as a badge suggesting that the naturally occurring radioactive carbon, C^{14}, and potassium, K^{40}, in our bodies "keep us healthy."

Risk assessment

The use of nuclear reactions as a source of energy has potential benefits, and potential dangers. Were it not for the unprecedented scale of both, the balance between the two would normally be assigned to a growing discipline called, "risk management."

Not only insurance companies, but many industries and municipal governments employ "risk managers." Their task is to evaluate risks, so that the policy makers can make appropriate decisions.

Building homes on the islands off the shores of North Carolina is a risk that some have taken. Because of the particular vulnerability to hurricanes every season, these homes are virtually uninsurable. Those who build there weigh the risk, and consider the periodic cost of rebuilding them an "operating expense." On balance, they decide to accept the periodic losses as part of the cost of what they regard as a paradise summer home.

The calculation of that risk is based on many years' experience; the losses are, statistically, predictable. As are the risks taken by auto insurance companies, which can know in advance how many and what kind of accidents they are insuring motorists for, based on the data from past years.

Municipalities balance the savings of imperfectly maintained streets against the risk of liability suits brought by the victims of accidents that can be attributed to negligent maintenance. Some risks are taken because the cost of avoiding them isn't worth it;

some are not taken because the cost of taking them is predictably too high.

The high school student who plays ice hockey knows that there is risk of injury, some of it within his control, some outside it. For every hundred who engage in this sport, there are a few for whom it was a bad risk, and then there are many for whom the same risk assessment turned out well. It is in the nature of risk that there are winners and losers. But all who take the risk are prepared for either.

The statistics of things that have never happened

There are two main differences between the kind of risk assessment described in the previous section, and the application of that discipline to nuclear risks, because the stakes are incredibly high and because there is no statistical basis for making the assessments.

There is generally a different kind of risk decision made. In both the weapons area and the arena of reactors, every possible precaution is taken, because the disaster that is being risked is so great that it must have no chance of ever happening.

This is, of course, unrealistic. It is cost-prohibitive to take *every possible precaution*, although often those words are used. Every *affordable* precaution is taken. And even if every possible precaution were taken, it is not certain that there would be zero chance of a disastrous accident.

Nuclear accidents of various sorts have in fact been assigned probabilities. A nuclear accident costing 3000 lives is said to have a likelihood of occurring once in a hundred million reactor years.[2]

---

[2] "Nuclear Power: Both Sides", M Kaku and J. Trainer, Norton, NY 1982

Other estimates give a likelihood of a Three-Mile Island level accident every 10,000 reactor years.

Such assessments vary widely with who is making them. There is no agreement about how to assign probability based on evidence that is not statistically valid because it consists of samples of zero, or one or two actual events. One particular kind of mishap that was given a likelihood of happening once in 500 reactor years, occurred twice in unrelated incidents in the six months following this assessment.

This is the nature of the statistics of things that have never happened.

People who are asked to take calculated risks based on this kind of statistical assessment respond in varying ways. Different philosophies and attitudes lead different people to advocate different solutions and different public policies. For some, the magnitude of the prospect of nuclear war, or of a Chernobyl type accident, looms so large that they want no part of the risk, no matter how small. For some, risk is part of life, and they find the risk that a dam may break or that a stadium full of sports fans may collapse, on a par with the risk they perceive of nuclear disaster.

What can happen? - The "China syndrome"

What can happen in nuclear warfare is amply evident. The destruction of all intelligent life, or perhaps all life entirely, is easily within the capacity of the nuclear arsenals available and ready to launch on command. No one has argued that nuclear war is impossible. It is a subject for speculation whether a nuclear war can be a "limited war," that is, one which is stopped by mutual agreement after two or six or twenty cities have been wiped out, leaving survivors in a world which may be dark and cold for years or decades, but which can ultimately be rebuilt.

Accidents occurring in nuclear reactors are more limited, even in the worst case. It is true that a uranium reactor can not become a nuclear bomb. A fission bomb requires nuclear fuel that is greatly enriched; it is doubtful that anything short of 90% enrichment is sufficient. Uranium reactors use 3% enriched uranium; graphite and heavy water reactors can operate with natural (0.7%) un-enriched uranium. Plutonium reactors, although they can become enriched beyond the original fraction, generally contain a sufficient proportion of un-converted U^{238} to prevent a "bomb" type accident.

This does not mean that a reactor can not explode, scattering its content of already fissioned but now radioactive material, as it did in Chernobyl.

The greatest danger in reactors comes from the loss of the ability to remove the heat produced in the core. SCRAM is a general remedy that stops fission instantly. There is, of course, no absolute guarantee that the control rods will be released, or that, if released, they will fall. One gets accustomed to considering eventualities that are outside the realm of planned-for contingencies. But because the release of the control rods occurs by default, if power is shut down, SCRAM is one of the more reliable safety features.

SCRAM stops fission within a second, but in any reactor that has been in operation for some time, heat continues to be produced in the fuel rods by the radioactive dissociation of the fission products. There is no way to stop that process – radioactivity can not be turned off. The rate at which heat is produced due to this factor starts at 7% of the rate of the reactor's energy production just before shut-down; it decreases in about a minute to 4%, in five minutes to 3%, in 15 minutes to 2%, but then goes for three hours before declining to 1%, a day to ½%, 11 days to ¼%, and does not stop altogether for thousands of years.

These may appear to be small percentages, but even 1% of the normal heat produced in the core, if not removed, is a large rate of energy production.

If there is leakage of the primary coolant, or if steam displaces water in the core, or for any number of other reasons, the core can become "uncovered," meaning that all this heat goes into the uranium of the fuel rods, eventually melting it.

The fuel rods tend to melt from the top down. If the entire core melts, there will be a puddle consisting of hundreds or thousands of tons of molten uranium, still containing the radioactive fission products which continue to heat it, at the bottom of the reactor vessel. The ultimate consequence is the melting of the iron vessel-bottom, with the uranium flowing to the bottom of the reactor building. There is general agreement of what would happen up to this point.

Given a high enough temperature, even concrete will melt, and this heated metal would go through the floor to the ground outside it. The name given to this eventuality is "meltdown." Some writers claim that this would not happen.

If it does, the moisture in the ground would be quickly vaporized into steam at high pressure, which will force the mix of clay and rock and radioactive fission products to rise around the building's foundation and break out into the atmosphere.

The image of melted uranium getting hotter and hotter, melting the ground downward all the way through the earth to China, has given rise to the term, "China syndrome." This will not actually occur, of course, but the name is applied to the more plausible outcome of a meltdown, the explosion of ground water boiling into a mix of high pressure steam and radioactive spent reactor fuel, going into the atmosphere and spreading far and wide.

No reactor has ever melted down.

By far the most devastating nuclear reactor accident was the one in which the top blew off a reactor building in Chernobyl in 1986. A cloud of core material spread thousands of miles and dropped radioactive rain that affected over 200,000,000 people, directly, through the air they breathed, through the food they ate, the milk they drank. What happened there is a story that has by now been fairly accurately documented. Technicians were conducting a test, during which normal safety devices had been purposely disabled. They misjudged alarm data, were ordered to continue in spite of signs that excess energy was being produced, and suddenly the top blew off the reactor, releasing much of the core material within.

Three Mile Island

Of the major accidents involving reactors, the one we know best is the accident at Three Mile Island, an island in the Susquehanna River in eastern Pennsylvania, about ten miles southeast of Harrisburg, the state capital.

One of the recurring themes in the tales of reactor accidents is the sense that they need not have happened.

With "all possible precautions" being taken, both in the automatic controls, the back-ups for all vital parts always on the ready, and backups for the backups, and the thorough training and careful selection of the technicians in the control room, why, then did these accidents happen?

The assurance that "nothing like *this* can ever happen again," does not answer the question, "what is the next *unexpected* thing that will happen because we are not prepared for *it*?"

There are not clear answers that are convincing to everyone.

For some eighteen years, physics students at Huron High School in Ann Arbor have found it informative to hear (we read it out loud) a most extraordinary account that was contained as an attachment to the Report of the President's Commission on the Three Mile Island Accident.[3]

The account, quite unlike the government document to which it was attached, is in eminently readable, diary form, a minute by minute retelling of the events. Understandable to the non-expert, it deals with what the individuals involved experienced, and goes from the scientific and technical description of what went wrong with the reactor, to the wryly amusing account of how the story was broken to the public by the local radio station, WKBO, which had been alerted by its traffic reporter, known as Captain Dave, who heard on his CB radio that fire fighters and police were mobilizing in nearby Middletown.

The account raises in personal terms the breakdown of well rehearsed emergency procedures, and gave our students an opportunity to discuss after the reading, such questions as, "Whom would you blame here?" "Could something like this happen again?" "What would you do if you were the head of the Nuclear Regulatory Commission, or a U.S. Senator, or the President?"

Many interesting responses came from these seventeen-year-olds. By and large, they are much less forgiving or tolerant of the failures of both the individuals and the institutions that run and regulate nuclear power than are most authors who write about nuclear issues.

The account is attached as an Appendix to this book. It is not copyrighted, and may be reproduced without permission. It was originally a part of the Report of the President's Commission, published by the Government Printing Office.[4]

---

[3] The Commission was appointed by President Jimmy Carter shortly after the accident, and was chaired by George Kemeny, a distinguished physicist and Dartmouth College president.

[4] The Report of the President's Commission on the Three Mile Island Accident is out of print, but many research libraries have copies. The account is not bound with all the copies of the Report. The Pennsylvania State Library has copies that include the account.

The Report was published in October 1979, six months after the accident, and the account does not contain the astonishing discovery made only many years later, when it became safe to open the top of the reactor and lower video cameras toward the bottom, that the reactor had become almost 90% uncovered and was far closer to melting down completely than had originally been thought.

APPENDIX

Minute by minute,
Hour by hour

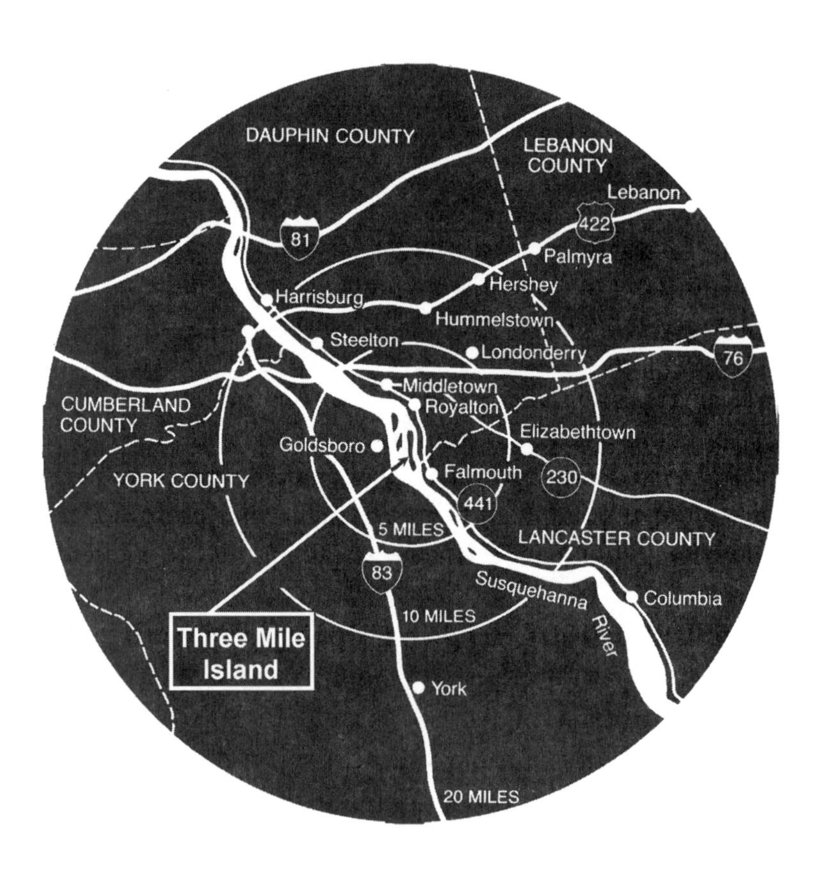

Fig A.1 Map of the area within 20 miles of Three Mile Island

Account of the Accident

What follows is remarkable because it takes you, minute by minute, hour by hour, day by day, to where the men and women who were caught up in the storm of this near-disaster were trying to cope with it.

It gives you a glimpse of what they saw and heard and what they knew and didn't, and how they made decisions that may have meant life or death for them and the hundreds of thousands of residents who lived within ten miles of the reactor at Three Mile Island, in eastern Pennsylvania, in the days following March 28, 1979.

This story is not just about Three Mile Island. It is about every nuclear reactor that ever was or ever will be, around the country and the world, providing electric power for homes and industry. Printing this account has no agenda, except to bring you as close as possible to the facts that might answer the question, "How could a reactor built to be safe under every foreseeable circumstance, have such a close call with catastrophe?"

You will find no simple answer, because probably there is none. Our students have gone into the reading of this account with some questions in mind, which we discussed afterward. You may wish to do the same.

What went wrong? Where was the line crossed between small incident and major accident?

Who were the major players, both before and during the days of this story who bear responsibility? Was anyone at fault? Who?
Are such accidents preventable? How?

What would you do if you were a member of the Nuclear Regulatory Commission? If you were a U.S. Senator or Representative? If you were President? If you were what you are: a citizen in a democratic country?

WEDNESDAY, MARCH 28

In the parlance of the electric power industry, a "trip" means a piece of machinery stops operating. A series of feedwater system pumps supplying water to TMI-2's steam generators tripped on the morning of March 28, 1979. The nuclear plant was operating at 97 percent power at the time. The first pump trip occurred at 36 seconds after 4:00 a.m. When the pumps stopped, the flow of water to the steam generators stopped. With no feedwater being added, there soon would be no steam, so the plant's safety system automatically shut down the steam turbine and the electric generator it powered. The incident at Three Mile Island was 2 seconds old.

The production of steam is a critical function of a nuclear reactor. Not only does steam run the generator to produce electricity but also, as steam is produced, it removes some of the intense heat that the reactor water carries.

When the feedwater flow stopped, the temperature of the reactor coolant increased. The rapidly heating water expanded. The pressurizer level (the level of the water inside the pressurizer tank) rose and the steam in the top of the tank compressed. Pressure inside the pressurizer built to 2,255 pounds per square inch, 100 psi more than normal. Then a valve atop the pressurizer, called a pilot-operated relief valve, or PORV, opened -- as it was designed to do -- and steam and water began flowing out of the reactor coolant system through a drain pipe to a tank on the floor of the containment building. Pressure continued to rise, however, and 8 seconds after the first pump tripped, TMI-2's reactor -- as it was designed to do -- scrammed: its control rods automatically dropped down into the reactor core to halt its nuclear fission.

Less than a second later, the heat generated by fission was essentially zero. But, as in any nuclear reactor, the decaying radioactive materials left from the fission process continued to heat the reactor's coolant water. This heat was a small fraction – just 6 percent – of that released during fission, but it was still substantial and had to be removed to keep the core from overheating. When the pumps that normally supply the steam generator with water shut down, three emergency feedwater pumps automatically started. Fourteen seconds into the accident, an operator in TMI-2's control room noted the emergency feed pumps were running. He did not notice two lights that told him a valve was closed on each of the two emergency feedwater lines and thus no water could reach the steam generators. One light was covered by a yellow maintenance tag. No one knows why the second light was missed.

With the reactor scrammed and the PORV open, pressure in the reactor coolant system fell. Up to this point, the reactor system was responding normally to a turbine trip. The PORV should have closed 13 seconds into the accident, when pressure dropped to 2,205psi. It did not. A light on the control room panel indicated that the electric power that opened the PORV had gone off, leading the operators to assume the valve had shut. But the PORV was stuck open, and would remain open for 2 hours and 22 minutes, draining needed coolant water -- a LOCA [Loss Of Coolant Accident] was in progress. In the first100 minutes of the accident, some 32,000 gallons -- over one-third of the entire capacity of the reactor coolant system -- would escape through the PORV and out the reactor's let-down system. Had the valve closed as it was designed to do, or if the control room operators had realized that the valve was stuck open and closed a backup valve to stem the flow of coolant water, or if they had simply left on the plant's high pressure injection pumps, the accident at Three Mile Island would have remained little more than a minor inconvenience for Met Ed.

***

To a casual visitor, the control room at TMI-2 can be an intimidating place, with messages coming from the loudspeaker of the plant's paging system; panel upon panel of red, green, amber, and white lights; and alarms that sound or flash warnings many times each hour. Reactor operators are trained how to respond and to respond quickly in emergencies. Initial actions are

Fig A.2 Control Room at Three Mile Island reactor

ingrained, almost automatic and unthinking. The burden of dealing with the early, crucial stages of the accident at Three Mile Island fell to four men -- William Zewe, shift supervisor in charge of both TMI-1 and TMI-2; Fred Scheimann, shift foreman for TMI-2; and two control room operators, Edward Frederick and Craig Faust. Each had been trained for his job by Met Ed and Babcock & Wilcox, the company that supplied TMI-2's reactor and nuclear steam system; each was licensed by the Nuclear Regulatory Commission; each was a product of his training – training that did not adequately prepare them to cope with the accident at TMI-2. Indeed, their training was partly responsible for escalating what should have been a minor event into a potentially devastating accident. Frederick and Faust were in the control room when the first alarm sounded, followed by a cascade of alarms that numbered 100 within minutes. The operators reacted quickly as trained to counter the turbine trip and reactor scram. Later Faust would recall for the Commission his reaction to the incessant alarms: "I would have liked to have thrown away the alarm panel. It wasn't giving us any useful information." Zewe, working in a small, glass-enclosed office behind the operators, alerted the TMI-1 control room of the TMI-2 scram and called his shift foreman back to the control room.

Scheimann had been overseeing maintenance on the plant's Number 7 polisher -- one of the machines that remove dissolved minerals from the feedwater system. His crew was using a mixture of air and water to break up resin that had clogged a resin transfer line. Later investigation would reveal that a faulty valve in one of the polishers allowed some water to leak into the air-controlled system that opens and closes the polishers' valves and may have been a factor in their sudden closure just before the accident began. This malfunction probably triggered the initial pump trip that led to the accident. The same problem of water leaking into the polishers' valve control system had occurred at least twice before at TMI-2. Had Met Ed corrected the earlier polisher problem, the March 28 sequence of events may never have begun.

***

With the PORV stuck open and heat being removed by the steam generators, the pressure and temperature of the reactor coolant system dropped. The water level also fell in the pressurizer. Thirteen seconds into the accident, the operators turned on a pump to add water to the system. This was done because the water in the system was shrinking as it cooled. Thus more water was needed to fill the system. Forty-eight seconds into the incident, while pressure continued falling, the water level in the pressurizer began to rise again. The reason, at this point, was that the amount of water being pumped into the system was greater than that being lost through the PORV.

About a minute and 45 seconds into the incident, because their emergency water lines were blocked, the steam generators boiled dry. After the steam generators boiled dry, the reactor coolant heated up again, expanded, and this helped send the pressurizer level up further.

Two minutes into the incident, with the pressurizer level still rising, pressure in the reactor coolant system dropped sharply. Automatically, two large pumps began pouring about 1,000 gallons a minute into the system. The pumps, called high pressure injection (HPI) pumps, are part of the reactor's emergency core cooling system. The level of water in the pressurizer continued to rise, and the operators, conditioned to maintain a certain level in the pressurizer, took this to mean that the system had plenty of water in it. However, the pressure of reactor coolant system water was falling, and its temperature became constant.

About 2½ minutes after the HPI pumps began working, Frederick shut one down and reduced the flow of the second to less than 100 gallons per minute. The falling pressure, coupled with a constant reactor coolant temperature after HPI came on, should have clearly alerted the operators that TMI-2 had suffered a LOCA, and safety required they maintain High Pressure Injection. "The rapidly increasing pressurizer level at the onset of the accident led me to believe that the high pressure injection was excessive, and that we were soon going to have a solid system," Frederick later told the Commission.

A solid system is one in which the entire reactor and its cooling system, including the pressurizer, are filled with water. The operators had been taught to keep the system from "going solid" -a condition that would make controlling the pressure within the reactor coolant system more difficult and that might damage the system. The operators followed this line of reasoning, oblivious for over 4 hours to a far greater threat -- that the loss of water from the system could result in uncovering the core.

The saturation point was reached 5-1/2 minutes into the accident. Steam bubbles began forming in the reactor coolant system, displacing the coolant water in the reactor itself. The displaced water moved into the pressurizer, sending its level still higher. This continued to suggest to the operators that there was plenty of water in the system. They did not realize that water was actually flashing into steam in the reactor, and with more water leaving the system than being added, the core was on its way to being uncovered. And so the operators began draining off the reactor's cooling water through piping called the let-down system.

Eight minutes into the accident, someone -- just who is a matter of dispute -- discovered that no emergency feedwater was reaching the steam generators. Operator Faust scanned the lights on the control panel that indicate whether the emergency feedwater valves are open or closed. He first checked a set of emergency feedwater valves designed to open after the pumps reach full speed; they were open. Next he checked a second pair of emergency feedwater valves, called the "twelve-valves," which are always supposed to be open, except during a specific test of the emergency feedwater pumps. The two "twelve-valves" were closed. Faust opened them and water rushed into the steam generators.

Fig A.3 TMI-2 reactor had 36,816 fuel rods in 177 fuel rod assemblies. One assembly shown being lowered into the core.

The two "twelve-valves" were known to have been closed 2 days earlier, on March 26, as part of a routine test of the emergency feedwater pumps. A Commission investigation has not identified a specific reason as to why the valves were closed at 8 minutes into the accident. The most likely explanations are: the valves were never reopened after the March 26 test; or the valves were reopened and the control room operators mistakenly closed the valves during the very first part of the accident; or the valves were closed mistakenly from control points outside the control room after the test. The loss of emergency feedwater for 8 minutes had no significant effect on the outcome of the accident. But it did add to the confusion that distracted the operators as they sought to understand the cause of their primary problem.

–

Throughout the first 2 hours of the accident, the operators ignored or failed to recognize the significance of several things that should have warned them that they had an open PORV and a loss of-coolant accident. One was the high temperatures at the drain pipe that led from the PORV to the reactor coolant drain tank. One emergency procedure states that a pipe temperature of 200°F indicates an open PORV. Another states that when the drain pipe temperature reaches 130°F , the block valve beneath it should be closed. But the operators testified that the pipe temperature normally registered high because either the PORV or some other valve was leaking slightly. "I have seen, in reviewing logs since the accident, approximately 198 degrees," Zewe told the Commission. "But I can remember instances before . . . just over 200 degrees." So Zewe and his crew dismissed the significance of the temperature readings, which Zewe recalled as being in the 230°F range. Recorded data show the range reached 285°F. Zewe told the Commission that he regarded the high temperatures on the drain pipe as residual heat: ". . .Knowing that the relief valve had lifted, the downstream temperature I would expect to be high and that it would take some time for the pipe to cool down below the 200-degree set point."

***

At 4:11 a.m., an alarm signaled high water in the containment building's sump, a clear indication of a leak or break in the system. The water, mixed with steam, had come from the open PORV, first falling to the drain tank on the containment building floor and finally filling the tank and flowing into the sump. At 4:15 a.m., a rupture disc on the drain tank burst as pressure in the tank rose. This sent more slightly radioactive water onto the floor and into the sump. From the sump it was pumped to a tank in the nearby auxiliary building.

Five minutes later, at 4:20 a.m., instruments measuring the neutrons inside the core showed a count higher than normal, another indication -- unrecognized by the operators -that steam bubbles were present in the core and forcing cooling water away from the fuel rods. During this time, the temperature and pressure inside the containment building rose rapidly from the heat and steam escaping via the PORV and drain tank. The operators turned on the cooling equipment and fans inside the containment building. The fact that they failed to realize that these conditions resulted from a LOCA indicates a severe deficiency in their training to identify the symptoms of such an accident.

About this time, Edward Frederick took a call from the auxiliary building. He was told an instrument there indicated more than 6 feet of water in-the containment building sump. Frederick queried the control room computer and got the same answer. Frederick recommended shutting off the two sump pumps in the containment building. He did not know where the water was coming from and did not want to pump water of unknown origin, which might be radioactive, outside the containment building. Both sump pumps were stopped about 4:39 a.m. Before they were, however, as much as 8,000 gallons of slightly radioactive water may have been pumped into the auxiliary building. Only 39 minutes had passed since the start of the accident.

***

George Kunder, superintendent of technical support at TMI-2, arrived at the Island about 4:45 a.m., summoned by telephone. Kunder was duty officer that day, and he had been told TMI-2 had had a turbine trip and reactor scram. What he found upon his arrival was not what he expected. "I felt we were experiencing a very unusual situation, because I had never seen pressurizer level go high and peg in the high range, and at the same time, pressure being low," he told the Commission. "They have always performed consistently." Kunder's view was shared by the control room crew. They later described the accident as a combination of events they had never experienced, either in operating the plant or in their training simulations.

Fig A.4 Fuel rod assembly being lowered into core, seen from 50 feet up.

Shortly after 5:00 a.m., TMI-2's four reactor coolant pumps began vibrating severely. This resulted from pumping steam as well as water, and it was another indication that went unrecognized that the reactor's water was boiling into steam. The operators feared the violent shaking might damage the pumps -- which force water to circulate through the core -- or the coolant piping.

Zewe and his operators followed their training. At 5:14 a.m., two of the pumps were shut down. Twenty-seven minutes later, operators turned off the two remaining pumps, stopping the forced flow of cooling water through the core.

There was already evidence by approximately 6:00 a.m. that at least a few of the reactor's fuel rod claddings had ruptured from high gas pressures inside them, allowing some of the radioactive gases within the rods to escape into the coolant water. The early warning came from radiation alarms inside the containment building. With coolant continuing to stream out the open PORV and little water being added, the top of the core became uncovered and heated to the point where the zirconium alloy of the fuel rod cladding reacted with steam to produce hydrogen. Some of this hydrogen escaped into the containment building through the open PORV and drain tank; some of it remained within the reactor. This hydrogen, and possibly hydrogen produced later in the day, caused the explosion in the containment building on Wednesday afternoon and formed the gas bubble that produced such great concern a few days later.

Other TMI officials now were arriving in the TMI-2 control room. They included Richard Dubiel, a health physicist who served as supervisor of radiation protection and chemistry; Joseph Logan, superintendent of TMI-2; and Michael Ross, supervisor of operations for TMI-1.

Shortly after 6:00 a.m., George Kunder participated in a telephone conference call with John Herbein, Met Ed's vice president for generation; Gary Miller, TMI station manager and Met Ed's senior executive stationed at the nuclear facility;' and Leland Rogers, the Babcock & Wilcox site representative at TMI.

The four men discussed the situation at the plant. In his deposition, Rogers recalled a significant question he posed during that call: he asked if the block valve between the pressurizer and the PORV, a backup valve that could be closed if the PORV stuck open, had been shut.

QUESTION: What was the response?

ROGERS: George's immediate response was, "I don't know," and he had someone standing next to the shift supervisor over back of the control room and sent the guy to find out if the valve block was shut.

QUESTION: You heard him give these instructions?

ROGERS: Yes, and very shortly I heard the answer come back from the other person to George, and he said, "Yes, the block valve was shut . . . . "

The operators shut the block valve at 6:22 a.m., 2 hours and 22 minutes after the PORV had opened.

It remains, however, an open question whether Rogers or someone else was responsible for the valve being closed. Edward Frederick testified that the valve was closed at the suggestion of a shift supervisor coming onto the next shift; but Frederick has also testified that the valve was closed because he and his fellow operators could think of nothing else to do to bring the reactor back under control.

In any event, the loss of coolant was stopped, and pressure began to rise, but the damage continued. Evidence now indicates the water in the reactor was below the top of the core at 6:15 a.m. Yet for some unexplained reason, high pressure injection to replace the water lost through the PORV and let-down system was not initiated for almost another hour. Before that occurred, Kunder, Dubiel, and their colleagues would realize they faced a serious emergency at TMI-2.

In the 2 hours after the turbine trip, periodic alarms warned of low-level radiation within the unoccupied containment building. After 6:00 a.m., the radiation readings markedly increased. About 6:30 a.m., a radiation technician began surveying the TMI-2 auxiliary building, using a portable detector -- a task that took about 20 minutes. He reported rapidly increasing levels of radiation, up to one rem per hour. During this period, monitors in the containment and auxiliary buildings showed rising radiation levels. By 6:48 a.m., high radiation levels existed in several areas of the plant, and evidence indicates as much as two-thirds of the 12-foot high core stood uncovered at this time. Analyses and calculations made after the accident indicate temperatures as high as 3,500 to 4,000°F or more in parts of the core existed during its maximum uncovery. At 6:54 a.m., the operators turned on one of the reactor coolant pumps, but shut it down 19 minutes later because of high vibrations. More radiation alarms went off. Shortly before 7:00 a.m., Kunder and Zewe declared a <u>site emergency</u> required by TMI's emergency plan whenever some event threatens "an uncontrolled release of radioactivity to the immediate environment."

***

Gary Miller, TMI station manager, arrived at the TMI-2 control room a few minutes after 7:00 a.m. . Radiation levels were increasing throughout the Plant. Miller had first learned of the turbine trip and reactor scram within minutes after they occurred. He had had several telephone conversations with people at the site, including the 6:00 a.m. conference call. When he reached Three Mile Island, Miller found that a site emergency existed. He immediately assumed command-as emergency director and formed a team of senior employees to aid him in controlling the accident and in implementing TMI-2's emergency plan.

Miller told Michael Ross to supervise operator activities in the TMI-2 control room. Richard Dubiel directed radiation activities, including monitorings on- and off-site. Joseph Logan was charged with ensuring that all required procedures and plans were reviewed and followed. George Kunder took over technical support and communications. Daniel Shovlin, TMI's maintenance superintendent, directed emergency maintenance. B&W's Leland Rogers was asked to provide technical assistance and serve as liaison with his home office. Miller gave James Seelinger, superintendent of TMI-1, charge of the emergency control station set up in the TMI-1 control room. Under TMI's emergency plan, the control room of the unit not involved in an accident becomes the emergency control station. On March 28, TMI-1 was in the process of starting again after being shut down for refueling of its reactor.

TMI personnel were already following the emergency plan, telephoning state authorities about the site emergency. The Pennsylvania Emergency Management Agency (PEMA) was asked to notify the Bureau of Radiation Protection (BRP), part of Pennsylvania's Department of Environmental Resources. The bureau in turn telephoned Kevin Molloy, director of the Dauphin County Office of Emergency Preparedness. Dauphin County includes Harrisburg and Three Mile Island. Other nearby counties and the State Police were alerted.

Met Ed alerted the U.S. Department of Energy's Radiological Assistance Plan office at Brookhaven National Laboratory. But notifying the Nuclear Regulatory Commission's Region I office in King of Prussia, Pennsylvania, took longer. The initial phone call reached an answering service, which tried to telephone the

NRC duty officer and the region's deputy director at their homes. Both were en route to work.

By the time the NRC learned of the accident -- when its Region I office opened at 7:45 a.m. -- Miller had escalated the site emergency at Three Mile Island to a general emergency. Shortly after 7:15 a.m., emergency workers had to evacuate the TMI-2 auxiliary building. William Dornsife, a nuclear engineer with the Pennsylvania Bureau of Radiation Protection, was on the telephone to the TMI-2 control room at the time. He heard the evacuation ordered over the plant's paging system. "And I said to myself, 'This is the biggie,' " Dornsife recalled in his deposition.

At 7:20 a.m., an alarm indicated that the radiation dome monitor high in the containment building was reading 8 rems per hour. The monitor is shielded by lead. This shielding is designed to cut the radioactivity reaching the monitor by 100 times. Thus, those in the control room interpreted the monitor's alarm as meaning that radiation present in the containment building at the time was about 800 rems per hour. Almost simultaneously, the operators finally turned on the high pressure injection pumps, once again dumping water into the reactor, but this intense flow was kept on for only 18 minutes. Other radiation alarms sounded in the control room. Gary Miller declared a general emergency at 7:24 a.m. By definition, at Three Mile Island, a general emergency is an "incident which has the potential for serious radiological consequences to the health and safety of the general public."

As part of TMI's emergency plan, state authorities were again notified and teams were sent to monitor radiation on the Island and ashore. The first team, designated Alpha and consisting of two radiation technicians, was sent to the west side of the Island, the downwind direction at the time. Another two-man team, designated Charlie, left for Goldsboro, a community of some 600 persons on the west bank of the Susquehanna River across from Three Mile Island. Meanwhile, a team sent into the auxiliary building reported increasing radiation levels and the building's basement partly flooded with water. At 7:48 a.m., radiation team Alpha reported radiation levels along the Island's west shoreline were less than one millirem per hour. Minutes later, another radiation team reported similar readings at the Island's north gate and along Route 441, which runs parallel to the Susquehanna's eastern shore.

***

Nearly 4 hours after the accident began, the containment building automatically isolated. Isolation is intended to help prevent radioactive material released by an accident from escaping into the environment. The building is not totally closed off. Pipes carrying coolant run between the containment and auxiliary buildings. These pipes close off when the containment building isolates, but the operators can open them. This occurred at TMI-2 and radioactive water flowed through these pipes even during isolation. Some of this piping leaked radioactive material into the auxiliary building, some of which escaped from there into the atmosphere outside.

In September 1975, the NRC instituted its Standard Review Plan, which included new criteria for isolation. The plan listed three conditions -- increased pressure, rising radiation levels, and emergency core cooling system activation -- and required that containment buildings isolate on any two of the three. However, the plan was not applied to nuclear plants that had already received their construction permits. TMI-2 had, so it was "grandfathered" and not required to meet the Standard Review Plan, although the plant had yet to receive its operating license.

In the TMI-2 design, isolation occurred only when increasing pressure in the containment building reached a certain point, nominally 4 pounds per square inch. Radiation releases alone, no matter how intense, would not initiate isolation, nor would ECCS [Emergency Core Cooling System] activation.

Although large amounts of steam entered the containment building early in the TMI-2 accident through the open PORV, the operators had kept pressure there low by using the building's cooling and ventilation system. But the failure to isolate early made little difference in the TMI-2 accident. Some of the radioactivity ultimately released into the atmosphere occurred after, isolation from leaks in the let-down system that continued to carry radioactive water out of the containment building into the auxiliary building.

At 8:26 a.m., the operators once again turned on the ECCS's high pressure injection pumps and maintained a relatively high rate of flow. The core was still uncovered at this time and evidence indicates it took until about 10:30 a.m. for the HPI pumps to fully cover the core again.

By 7:50 a.m., NRC Region I officials had established direct telephone contact with the TMI-2 control room. Ten minutes later, Region I activated its Incident Response Center at King of Prussia, opened a direct telephone line to the Emergency Control Station in the TMI-1 control room, and notified NRC staff headquarters in Bethesda, Maryland. Region I officials gathered what information they could and relayed it to NRC headquarters, which had activated its own Incident Response Center. Region I dispatched two teams of inspectors to Three Mile Island; the first left at about 8:45 a.m., the second a few minutes later.

Around 8:00 a.m. it was clear to Gary Miller that the TMI-2 reactor had suffered some fuel damage. The radiation levels told him that. Yet Miller would testify to the Commission: ". . . I don't believe in my mind I really believed the core had been totally uncovered, or uncovered to a substantial degree at that time."

Off the Island, radiation readings continued to be encouragingly low. Survey team Charlie-reported no detectable radiation in Goldsboro. Miller and several aides concluded about 8:30 a.m. that the emergency plan was being properly implemented.

***

WKBO, a Harrisburg "Top 40" music station, broke the story of TMI-2 on its 8:25 a.m. newscast. The station's traffic reporter, known as Captain Dave, uses an automobile equipped with a CB radio to gather his information. About 8:00 a.m. he heard police and fire fighters were mobilizing in Middletown and relayed this to his station. Mike Pintek, WKBO's news director, called Three Mile Island and asked for a public relations official. He was connected instead with the control room to a man who told him: "I can't talk now, we've got a problem." The man denied that "there are any fire engines," and told Pintek to telephone Met Ed's headquarters in Reading, Pennsylvania.

Pintek did, and finally reached Blaine Fabian, the company's manager of communications services. In an interview with the Commission staff, Pintek told what happened next:

Fabian came on and said there was a general emergency. What the hell is that? He said that general emergency was a "red-tape" type of thing required by the NRC when certain conditions exist. What conditions? "There was a problem with a feedwater pump. The plant is shut down. We're working on it. There's no danger off-site. No danger to the general public." And that is the story we went with at 8:25. 1 tried to tone it down so people wouldn't be alarmed.

At 9:06 a.m., the Associated Press filed its first story -- a brief dispatch teletyped to newspaper, television, and radio news rooms across the nation. The article quoted Pennsylvania State Police as saying a general emergency had been declared, "there is no radiation leak," and that Met Ed officials had requested a State Police helicopter "that will carry a monitoring team." The story contained only six sentences in four paragraphs, but it alerted editors to what would become one of the most heavily reported news stories of 1979.

Many public officials learned of the accident from the news media, rather than from the state, or their own emergency preparedness people. Harrisburg Mayor Paul Doutrich was one, and that still rankled him when he testified before the Commission 7 weeks later. Doutrich heard about the problem in a 9:15 a.m. telephone call from a radio station in Boston. "They asked me what we were doing about the nuclear emergency," Doutrich recalled. "My response was, 'What nuclear emergency?' They said, 'Well, at Three Mile Island.' I said, 'I know nothing about it. We have a nuclear plant there, but I know nothing about a problem.' So they told me; a Boston radio station."

***

At 9:15 a.m., the NRC notified the White House of the events at Three Mile Island. Seven minutes later, an air sample taken in Goldsboro detected low levels of radioactive iodine-131. This specific reading was erroneous; a later, more sensitive analysis of the sample found no iodine-131. At 9:30 a.m., John Herbein, Met Ed's vice president for generation, was ordered to Three Mile Island from Philadelphia by Met Ed President Walter Creitz. And at 10:05 a.m., the first contingent of NRC Region I officials arrived at Three Mile Island.

In the days to follow, the NRC would dominate the public's perception of the events at Three Mile Island. But the initial NRC team consisted of only five Region I inspectors, headed by Charles Gallina. The five were briefed in the TMI-1 control room on the status of TMI-2. Then Gallina sent two inspectors into the TMI-2 control room and two more out to take radiation measurements; he himself remained in the TMI-1 control room to coordinate their reports and relay information to both Region I and NRC headquarters.

While the NRC team received its briefing, monitors indicated that radiation levels in the TMI-2 control room had risen above the levels considered acceptable in NRC regulations. Workers put on protective face masks with filters to screen out any airborne radioactive particles. This made communications among those managing the accident difficult. At 11:00 a.m., all nonessential personnel were ordered off the Island. At the same hour, both Pennsylvania's Bureau of Radiation Protection and the NRC requested the Department of Energy to send a team from Brookhaven National Laboratory to assist in monitoring environmental radiation.

About this time, Mayor Robert Reid of Middletown telephoned Met Ed's home office in Reading. He was assured, he later told the Commission, that no radioactive particles had escaped and no one was injured.

> I felt relieved and relaxed; I said, "There's no problem." Twenty seconds later I walked out of my office and got in my car and turned the radio on and the announcer told me, over the radio, that there were radioactive particles released. Now, I said, "Gee whiz, what's going on here?" At 4:00 in the

> afternoon the same day the same man called me at home and said, "Mayor Reid, I want to update our conversation that we had at 11:00 a.m." I said, "Are you going to tell me that [radioactive] particles were released?" He said,' "Yes." I said, "I knew that 20 seconds after I spoke to you on the phone. "

Throughout much of the morning, Pennsylvania's Lieutenant Governor William Scranton, III, focused his attention on Three Mile Island. Scranton was charged, among other things, with overseeing the state's emergency preparedness functions. He had planned a morning press conference on energy conservation, but when he finally faced reporters in Harrisburg, the subject was TMI-2. In a brief opening statement, Scranton said:

> The Metropolitan Edison Company has informed us that there has been an incident at Three Mile Island, Unit-2. Everything is under control. There is and was no danger to public health and safety. . . . There was a small release of radiation to the environment. All safety equipment functioned properly. Metropolitan Edison has been monitoring the air in the vicinity of the plant constantly since the incident. No increase in normal radiation levels has been detected . . . .

During the questioning by reporters, however, William Dornsife of the state's Bureau of Radiation Protection, who was there at Scranton's invitation, said Met Ed employees had "detected a small amount of radioactive iodine. . . ." Dornsife had learned of the iodine reading (later found to be in error) just before the press conference began and had not had time to tell Scranton. Dornsife dismissed any threat to human health from the amount of radioactive iodine reported in Goldsboro.

Shortly after the press conference, a reporter told Scranton that Met Ed in Reading denied any off-site radiation. While some company executives were acknowledging radiation readings off the Island, low-level public relations officials at Met Ed's headquarters continued until noon to deny any off-site releases. It was an error in communications within Met Ed, one of several that would reduce the utility's credibility with public officials and the press. "This was the first contradictory bit of information that we received and it caused some disturbance," Scranton told the Commission in his testimony.

***

At Three Mile Island, the control room was crowded with operator and supervisors trying to bring the plant under control. They had failed in efforts to establish natural circulation cooling. This essentially means setting up a flow of water, without mechanical assistance, by heating water in the core and cooling it in the steam generators. This effort failed because the reactor coolant system was not filled with water and a gas bubble forming in the top of the reactor blocked this flow of water. At 11:38 a.m., operators began to decrease pressure in the reactor system. The pressurizer block valve was opened and high pressure injection cut sharply. This resulted again in a loss of coolant and an uncovering of the core. The depressurization attempt ended at 3:08 p.m. The amount and duration of core uncovery during this period remains unknown.

Fig A.5 Inspection tags on control room panel. One of these had covered a signal light.

About noon, three employees entered the auxiliary building and found radiation levels ranging from 50 millirems to 1,000 rems (one million millirems) an hour. Each of the three workers received an 800-millirem dose during the entry. At 12:45 p.m., the Pennsylvania State Police closed Route 441 to traffic near Three Mile Island at the request of the state's Bureau of Radiation Protection. An hour later, the U.S. Department of Energy team began its first helicopter flight to monitor radiation levels. And at 1:50 p.m., a noise penetrated the TMI-2 control room; "a thud," as Gary Miller later characterized it.

That thud was the sound of a hydrogen explosion inside the containment building. It was heard in the control room; its force of 28 pounds per square inch was recorded on a computer strip chart there, which Met Ed's Michael Ross examined within a minute or two. Yet Ross and others failed to realize the significance of the event. Not until late Thursday was that sudden and brief rise in pressure recognized as an explosion of hydrogen gas released from the reactor. The noise, said B&W's Leland Rogers in his deposition, was dismissed at the time as the slamming of a ventilation damper. And the pressure spike on the strip chart, Ross explained to the Commission, "we kind of wrote it off . . . [as] possibly instrument malfunction. . . ."

Miller, Herbein, and Kunder left for Harrisburg soon afterwards for a 2:30 p.m. briefing with Lieutenant Governor Scranton on the events at Three Mile Island. At 2:27 p.m., radiation readings in Middletown ranged from I to 2 millirems per hour.

***

The influx of news media from outside the Harrisburg area began during the afternoon. The wire service reports of Associated Press and United Press International had alerted editors here and abroad to the accident. The heavy flow of newspaper and magazine reporters, television and radio correspondents, and photographers and camera crews would come later as the sense of concern about Three Mile Island grew. But at 4:30 p.m., when Scranton once more met the press, he found some strange faces among the familiar crew of correspondents who regularly covered Pennsylvania's Capitol.

Scranton had discussed the TMI situation with his own people and listened to Met Ed officials. "I wouldn't say that they [Met Ed] were exactly helpful, but they were not obstructive," he later testified. "I think they were defensive." Scranton was disturbed by, among other things, Herbein's comment during their 2:30 p.m. meeting that Herbein had not told reporters about some radiation releases during an earlier Met Ed press conference because "it didn't come up." So Scranton was less assured about conditions at Three Mile Island when he issued his afternoon statement to the press:

This situation is more complex than the company first led us to believe. We are taking more tests. And at this point, we believe there is still no danger to public health. Metropolitan Edison has given you and us conflicting information. We just concluded a meeting with company officials and hope this briefing will clear up most of your questions. There has been a release of radioactivity into the environment. The magnitude of the release is still being determined, but there is no evidence yet that it has resulted in the presence of dangerous levels. The company has informed us that from 11:00 a.m. until about 1:30 p.m., Three Mile Island discharged into the air, steam that contained detectable amounts of radiation. . . .

Scranton's statement inappropriately focused public attention on the steam emissions from TMI-2 as a source of radiation. In fact, they were not, since the water that flows inside the towers is in a closed loop and cannot mix with water containing radioactive materials unless there is a leak in the system.

Scranton went on to discuss potential health effects of the radiation releases:

> The levels that were detected were below any existing or proposed emergency action levels. But we are concerned because any increased exposure carries with it some increased health risks. The full impact on public health is being evaluated as environmental samples are analyzed. We are concerned most about radioactive iodine, which can accumulate in the thyroid, either through breathing or through drinking milk. Fortunately, we don't believe the risk is significant because most dairy cows are on stored feed at this time of year.

Many Americans learned about the accident at Three Mile Island from the evening newscasts of the television networks. Millions, for example, watched as Walter Cronkite led off the CBS Evening News:

> It was the first step in a nuclear nightmare; as far as we know at this hour, no worse than that. But a government official said that a breakdown in an atomic power plant in Pennsylvania today is probably the worst nuclear accident to date . . .

At 7:30 p.m., Mayor Ken Myers of Goldsboro met with the borough council to discuss the accident and the borough's evacuation plan. Then Myers suggested he and the council members go door-to-door to talk with residents of the small community.

> Everyone listened to what we had to say. We mainly told them of what we had heard through the radio, TV, and even our own public relations and communications department in the basement of the York County court house. . . Then we

told them also of our evacuation plans in case the Governor would declare an emergency and that we would all have to leave. Of course, right away they gave us questions: "Well, what should we do? Do you think it's safe that we should stay or do you think we should go?" The ones that I talked to, I told them: "Use your own judgment. We dare not tell you to leave your homes."

THURSDAY, MARCH 29

In retrospect, Thursday seemed a day of calm. A sense of betterment, if not well-being, was the spirit for much of the day. Radiation levels remained high at points within the auxiliary building, but off-site readings indicated no problems. The log book kept by the Dauphin County Office of Emergency Preparedness reflects this mood of a crisis passing:

5:45 a.m.	Called Pennsylvania Emergency Management Agency -- Blaisdale, reactor remains under control more stable than yesterday, not back to normal, monitoring continues by Met Ed, Radiological Health, and Nuclear Regulatory Commission.
7:55 a.m.	Pennsylvania Emergency Management Agency -- ... no danger to public.
11:25 a.m.	Pennsylvania Emergency Management Agency advised situation same.
3:30 p.m.	. . . situation is improving.
6:12 p.m.	. . . no change -- not cold yet, continues to improve, slow rate, off-site release controlled.
7:00-9:00 p.m.	. . . Pennsylvania Emergency Management says Island getting better.
9:55 p.m.	. . . no real measurable reading off-site -- no health risk off-site, no emergency, bringing reactor to cold shut down. . . .

Radiation monitoring continued. Midmorning readings showed 5 to 10 millirems an hour on-site and 1 to 3 millirems per hour across the Susquehanna River to the west. No radioactive iodine was detected in the air. The U.S. Food and Drug Administration began monitoring food, milk, and water in the area for radiation contamination.

Thursday was a day of questioning. NRC Chairman Joseph Hendrie and several key aides journeyed to Capitol Hill to brief the House Subcommittee on Energy and the Environment and other members of Congress on the accident. Lieutenant Governor Scranton spent several hours in the early afternoon at Three Mile Island, touring the TMI-2 control room and auxiliary building, wearing a radiation suit and respirator during part of his inspection. That same afternoon, Met Ed officials and NRC inspectors briefed several visiting members of Congress, including Rep. Allen Ertel (D-Pa.), whose district includes Three Mile Island, and Sen. John Heinz (R-Pa.). Later in the day, a second Congressional delegation that included Sen. Richard Schweiker (R-Pa.) and Rep. William Goodling (R-Pa.), whose district includes York, Adams, and Cumberland counties, received a briefing.

Thursday was also a day of disquieting discussions and discoveries. Thursday afternoon, a telephone conversation took place between two old acquaintances, Gordon MacLeod, Pennsylvania's Secretary of Health, and Anthony Robbins, director of the National Institute for Occupational Safety and Health. One important point of that conversation remains in dispute. MacLeod recalls that Robbins urged him to recommend an evacuation Robbins denies discussing or suggesting such an evacuation.

Up to this point, MacLeod -- who had taken office only 12 days before the accident -- had offered no recommendations since his department had no direct responsibility for radiological health matters. Now, however, he arranged a conference telephone call with Oran Henderson, director of the

Pennsylvania Emergency Management Agency; Thomas Gerusky, director of the Bureau of Radiation Protection; and John Pierce, an aide to Lieutenant Governor Scranton. MacLeod told them Robbins had strongly recommended evacuation. The others rejected the idea, although they agreed it should be reconsidered if conditions proved worse than they appeared at TMI-2. MacLeod then asked if it might be wise to have pregnant women and children under age 2 leave the area around the nuclear plant. This, too, was rejected Thursday afternoon.

Fig A.6 A helicopter samples air above TMI-2 containment building

At 2:10 p.m., a helicopter over TMI-2 detected a brief burst of radiation that measured 3,000 millirems per hour 15 feet above the plant's vent. This information was relayed to NRC headquarters, where it created no great concern.

But another release that afternoon, one within NRC limits for radiation releases, did cause considerable consternation. Soon after the accident began Wednesday, Met Ed stopped discharging wastewater from such sources as toilets, showers, laundry facilities, and leakage in the turbine and control and service buildings into the Susquehanna River. Normally, this water contains little or no radioactivity, but as a result of the accident, some radioactive gases had contaminated it. The radiation levels, however, were within the limits set by the NRC. By Thursday afternoon, nearly 400,000 gallons of this slightly radioactive water had accumulated and the tanks were now close to overflowing. Two NRC officials – Charles Gallina on-site and George Smith at the Region I office -- told Met Ed they had no objections to releasing the water so long as it was within NRC specifications. Met Ed notified the Bureau of Radiation Protection and began dumping the wastewater. No communities downstream from the plant were informed, nor was the press.

When NRC Chairman Hendrie learned of the release, he ordered it stopped. Hendrie did not know the water's source, and he was concerned about the impact on the public of the release of any radiation, no matter how slight. Some 40,000 gallons had entered the river when the dumping ceased around 6:00 p.m. Both NRC officials on-site and the Governor's aides realized that authorizing release of the wastewater would be unpopular, and neither was eager to do so. Yet the tanks still were close to overflowing. After hours of discussion, agreement was reached on the wording of a press release that the state's Department of Environmental Resources issued, which said DER "reluctantly agrees that the action must be taken." Release of the wastewater resumed shortly after midnight.

Late Thursday afternoon, Governor Thornburgh had held a press conference. At it, the NRC's Charles Gallina told reporters the danger was over for people off the Island. Thornburgh distrusted the statement at the time, and events soon confirmed his suspicion. At 6:30 p.m., Gallina and James Higgins, an NRC reactor inspector, received the results of an analysis of the reactor's coolant water. It showed that core damage was far more substantial than either had anticipated[1]. At 10:00 p.m., Higgins telephoned the Governor's office with the new information and indicated that a greater possibility of radiation releases existed. Nothing had changed inside the plant, only NRC's awareness of the seriousness of the damage. Yet Higgins' call foretold events only hours away.

---

[1] Not until several years later, when it became possible to lower a video camera into the core was the full extent of the core damage known. Close to 90% of the core had melted.

FRIDAY, MARCH 30

The TMI-2 reactor has a means of removing water from the reactor coolant system, called the let-down system, and one for adding water, called the make-up system. Piping from both runs through the TMI-2 auxiliary building, and NRC officials suspected that leaks in these two systems explained the sporadic, uncontrolled releases of radioactivity. They were also concerned about levels in the make-up tank and the two waste gas decay tanks inside the auxiliary building. Water from the let-down system flows into the make-up tank. In that tank, gases dissolved in the reactor's cooling water at high pressure released because the tank's pressure is lower, much as the gas bubbles in a pressurized carbonated beverage appear when the bottle is opened. These gases, under normal circumstances, are compressed and stored in the waste gas decay tanks. NRC officials worried that if the waste gas decay tanks filled to capacity, relief valves would open, allowing a continuing escape of radiation into the environment. That concern and what Commission Chairman Kemeny would later call a "horrible coincidence" resulted in a morning of confusion, contradictory evacuation recommendations, and eventually an evacuation advisory from Governor Richard Thornburgh.

About halfway through his midnight-to-noon shift on Friday, James Floyd, TMI-2's supervisor of operations, decided to transfer radioactive gases from the make-up tank to a waste gas decay tank. Floyd knew this would release radiation because of leaks in the system, but he considered the transfer necessary. The pressure in the make-up tank was so high that water that normally flowed into it for transfer to the reactor coolant system could not enter the tank. Floyd, without checking with other TMI and Met Ed officials, ordered the transfer to begin at 7:10 a.m. to reduce the tank's pressure. This controlled release allowed radioactive material to escape into the auxiliary building and then into the air outside. Thirty-four minutes later, Floyd requested a helicopter be sent to take radiation measurements. The chopper reported readings of 1,000 millirems per hour at 7:56 a.m. and 1,200 millirems per hour at 8:01 a.m., 130 feet above the TMI-2 vent stack.

At NRC headquarters, Lake Barrett, a section leader in the environmental evaluation branch, was concerned about the waste gas decay tank level. The previous evening, he had helped calculate "a hypothetical release rate" for the radiation that would escape if the tank's relief valves opened. Shortly before 9:00 a.m., Barrett was told of a report from Three Mile Island that the waste gas decay tanks had filled. He was asked to brief senior NRC staff officials on the significance of this. The group included Lee Gossick, executive director for operations; John Davis, then acting director of Inspection and Enforcement; Harold Denton, director of Nuclear Reactor Regulation; Victor Stello, Jr., then director of the Office of Operating Reactors; and Harold Collins, assistant director for emergency preparedness in the Office of State Programs. During the briefing, Barrett was asked what the release rate would mean in terms of an off-site dose. He did a quick calculation and came up with a figure: 1,200 millirems per hour at ground level. Almost at that moment, someone in the room reported a reading of 1,200 millirems per hour had been detected at Three Mile Island. By coincidence, the reading from TMI was identical to the number calculated by Barrett. "It was the exact same number, and it was within maybe 10 or 15 seconds from my first 1,200 millirems per hour prediction," Barrett told the Commission.

The result was instant concern among the NRC officials; "an atmosphere of significant apprehension," as Collins described it in his testimony. Communications between the NRC headquarters and Three Mile Island had been less than satisfactory from the beginning. "I think there was uncertainty in the operations center as to precisely what was going on at the facility and the question was being raised in the minds of many as to whether or not those people up there would do the right thing at the right time, if it had to be done," Collins testified. NRC officials proceeded without confirming the reading and without knowing whether the 1,200 millirem per hour reading was on- or off-site, whether it was taken at ground level or from a helicopter, or what its source was. They would later learn that the radiation released did not come from the waste gas decay tanks. The report that these tanks had filled was in error.

After some discussion, Harold Denton directed Collins to notify Pennsylvania authorities that senior NRC officials recommended the Governor order an evacuation. Collins telephoned Oran Henderson, director of the Pennsylvania Emergency Management Agency, and, apparently selecting the distance on his own, recommended an evacuation of people as far as 10 miles downwind from Three Mile Island. Henderson telephoned Lieutenant Governor Scranton, who promised to call the Governor. A Henderson aide also notified Thomas Gerusky, director of the Bureau of Radiation Protection, of the evacuation recommendation. Gerusky knew of the 1,200 millirem reading. A telephone call to an NRC official at the plant reinforced Gerusky's belief that an evacuation was unnecessary. He tried to telephone Governor Thornburgh, found the lines busy, and went to the Governor's office to argue personally against an evacuation.

Kevin Molloy, director of emergency preparedness for Dauphin County, had received a call from Met Ed's James Floyd at 8:34 a.m., alerting him to the radiation release. Twenty minutes later, the Pennsylvania Emergency Management Agency notified Molloy of an on-site emergency and an increase in radiation, but Molloy was told that no evacuation was needed. Then at 9:25 a.m., Henderson called Molloy and told him to expect an official evacuation order in 5 minutes; the emergency preparedness offices in York and Lancaster counties received similar alerts. Molloy began his preparations. He notified all fire departments within 10 miles of the stricken plant, and broadcast a warning over radio station WHP that an evacuation might be called.

At Three Mile Island, NRC's Charles Gallina was confronted by a visibly upset Met Ed employee shortly after Molloy's broadcast. "As the best I can remember, he said, 'What the hell are you fellows doing? My wife just heard the NRC recommended evacuation,'" Gallina told the Commission. Gallina checked radiation readings on- and off-site and talked with an NRC reactor inspector, who said "things were getting better." Then Gallina telephoned NRC officials at Region I and at Bethesda headquarters in an attempt "to call back that evacuation notice."

Shortly after 10:00 a.m., Governor Thornburgh talked by telephone with Joseph Hendrie. The NRC chairman assured the Governor that no evacuation was needed. Still, Hendrie had a suggestion: that Thornburgh urge everyone within 5 miles downwind of the plant to stay indoors for the next half-hour. The Governor agreed and later that morning issued an advisory that all persons within 10 miles of the plant stay inside. During this conversation, Thornburgh asked Hendrie to send a single expert to Three Mile Island upon whom the Governor could rely for technical information and advice.

About an hour later, Thornburgh received a telephone call from President Carter, who had just talked with Hendrie. The President said that he would send the expert the Governor wanted. That expert would be Harold Denton. The President also promised that a special communications system would be set up to link Three Mile Island, the Governor's office, the White House, and the NRC.

Thornburgh convened a meeting of key aides to discuss conditions at Three Mile Island. During this meeting, at about 11:40 a.m., Hendrie again called the Governor. As Gerusky recalls the conversation that took place over a speaker phone, the NRC chairman apologized for the NRC staff error in recommending evacuation. Just before the call, Emmett Welch, an aide to Gordon MacLeod, had renewed the Secretary of Health's recommendation that pregnant women and children under age 2 be evacuated. Thornburgh told Hendrie of this. Gerusky recalls this response from Hendrie: "If my wife were pregnant and I had small children in the area, I would get them out because we don't know what is going to happen." After the call, Thornburgh decided to recommend that pregnant women and preschool children leave the region within a 5-mile radius of Three Mile Island and to close all schools within that area. He issued his advisory shortly after 12:30 p.m.

Thornburgh was conscious throughout the accident that an evacuation might be necessary, and this weighed upon him. He later shared some of his concerns in testimony before the Commission:

> There are known risks, I was told, in an evacuation. The movement of elderly persons, people in intensive care units, babies in incubators, the simple traffic on the highways that results from even the best of an orderly evacuation, are going to exert a toll in lives and injuries. Moreover, this type of evacuation had never been carried out before on the face of this earth, and it is an evacuation that was quite different in kind and quality than one undertaken in time of flood or hurricane or tornado . . . . When you talk about evacuating people within a 5-mile radius of the site of a nuclear reactor, you must recognize that that will have 10-mile consequences, 20-mile consequences, 100-mile consequences, as we heard during

> the course of this event. This is to say, it is an event that people are not able to see, to hear, to taste, to smell . . . .

***

Relations between reporters and Met Ed officials had deteriorated over several days. Many reporters suspected the company of providing them with erroneous information at best, or of outright lying. When John Herbein arrived at 11:00 a.m. Friday to brief reporters gathered at the American Legion Hall in Middletown, the situation worsened. The press corps knew that the radioactivity released earlier had been reported at 1,200 millirems per hour; Herbein did not. He opened his remarks by stating that the release had been measured at around 300 to 350 millirems per hour by an aircraft flying over the Island. The question-and-answer period that followed focused on the radiation reading -- "I hadn't heard the number 1,200," Herbein protested during the news conference -- whether the release was controlled or uncontrolled, and the previous dumping of radioactive wastewater. At one point Herbein said, "I don't know why we need to . . . tell you each and every thing that we do specifically. . . . " It was that remark that essentially eliminated any credibility Herbein and Met Ed had left with the press.

The next day, Jack Watson, a senior White House aide, would telephone Herman Dieckamp, president of Met Ed's parent company, to express his concern that the many conflicting statements about TMI-2 reported by the news media were increasing public anxiety. Watson would suggest that Denton alone brief reporters on the technical aspects of the accident and Dieckamp would agree.

***

The radiation release, Molloy's announcement of a probable evacuation, and finally the Governor's advisory brought concern and even fear to many residents. Some people had already left, quietly evacuating on their own; others now departed. "On March 29 of this year, my wife and I joyously brought home our second daughter from the hospital; she was just 6 days old," V.T. Smith told the Commission. "On the morning of the 30th, all hell broke loose and we left for Delaware to stay with relatives." By Saturday evening, a Goldsboro councilman estimated 90 percept of his community's residents had left.

Schools closed after the Governor's advisory. Pennsylvania State University called off classes for a week at its Middletown campus. Friday afternoon, " . . . still having heard nothing from three Mile Island," Harrisburg Mayor Paul Doutrich drove with his deputy public works director to the TMI Observation Center overlooking the nuclear facility. There they talked for an hour with Met Ed President Creitz and Vice President Herbein. "Oddly enough, one of the things that impressed me the most and gave me the most feeling of confidence that things were all right was that everybody in that area, all the employees, the president and so forth, were walking around in their shirt sleeves, bare-headed," Doutrich told the Commission. "I saw not one indication of nuclear protection."

Friday, Saturday, and Sunday were hectic days in the emergency preparedness offices of the counties close to Three Mile Island. Officials labored first to prepare 10-mile evacuation plans and then ones covering areas out to 20 miles from the plant. The Pennsylvania Emergency Management Agency recommended Friday morning that 10-mile plans be readied. The three counties closest to the nuclear plant already had plans to evacuate their residents -- a total of about 25,000 living within 5 miles of the Island. A 10-mile evacuation had never been contemplated. For Kevin Molloy in Dauphin County, extending the evacuation zone meant the involvement of several hospitals -

- something he had not confronted earlier. There were no hospitals within 5 miles. Late Friday night, PEMA told county officials to develop 20-mile plans. Suddenly, six counties were involved in planning for the evacuation of 650,000 people, 13 hospitals, and a prison.

Friday was also the day the nuclear industry became deeply involved in the accident. After the radiation release that morning, GPU President Dieckamp set about assembling an industry team to advise him in managing the emergency. Dieckamp and an aide talked with industry leaders around the country, outlining the skills and knowledge needed at TMI-2. By late Saturday afternoon, the first members of the Industry Advisory Group had arrived. They met with Dieckamp, identified the tasks that needed immediate attention, and decided who would work on each.

***

Harold Denton arrived on site about 2:00 p.m. Friday, bringing with him a cadre of a dozen or so experts from NRC headquarters. Earlier in the day, NRC had learned of the hydrogen burn or explosion that flashed through the containment building Wednesday afternoon. The NRC staff already knew that some form of gas bubble existed within the reactor system. Now it became obvious that the bubble, an estimated 1,000 cubic feet of gases, contained hydrogen. And as Denton would later recall in his deposition, the question arose whether there was a potential for a hydrogen explosion. Throughout Friday, Denton operated on estimates provided him before he left Bethesda, which indicated that the bubble could not self-ignite for 5 to 8 days. Denton focused his immediate attention on finding ways to eliminate the bubble.

At about 8:30 p.m. Friday, Denton briefed Governor Thornburgh in person for the first time. Fuel damage was extensive; the bubble posed a problem in cooling the core; no immediate evacuation was necessary, Denton said. Then the two men held their first joint press conference. The Governor reiterated that no evacuation was needed, lifted his advisory that people living within 10 miles of Three Mile Island stay indoors, but continued his recommendation that pregnant women and preschool children remain more than 5 miles from the plant.

***

Shortly after 4:00 p.m., Jack Watson, President Carter's assistant for intergovernmental affairs, called Jay Waldman, Governor Thornburgh's executive assistant. The two disagree about the substance of that call. In an interview with the Commission staff, Waldman said Watson asked that the Governor not request President Carter to declare a state of emergency or disaster:

> He said that it was their belief that that would generate unnecessary panic, that the mere statement that the President has declared this area an emergency and disaster area would trigger a substantial panic; and he assured me that we were getting every type and level of federal assistance that we would get if there had been a declaration. I told him that I would have to have his word on that, an absolute assurance, and that if that were true, I would go to the Governor with his request that we not formally ask for a declaration.

Watson and his assistant, Eugene Eidenberg, both said in their Commission depositions that the White House never asked Governor Thornburgh not to request such a declaration. Whatever was said in that Friday conversation, the Governor made no request to the President for an emergency declaration. State officials later expressed satisfaction with the assistance provided by the federal government during the accident and immediately after. They were less satisfied, however, in August with the degree of assistance and cooperation they were receiving from federal agencies.

***

Officials of the U.S. Department of Health, Education, and Welfare (HEW) had become concerned about the possible release of radioactive iodine at Three Mile Island and began Friday to search for potassium iodide -- a drug capable of preventing radioactive iodine from lodging in the thyroid. The thyroid absorbs potassium iodide to a level where the gland can hold no more. Thus, if a person is exposed to radioactive iodine after receiving a sufficient quantity of potassium iodide, the thyroid is saturated and cannot absorb the additional iodine with its potentially damaging radiation. At the time of the TMI-2 accident, however, no pharmaceutical or chemical company was marketing medical-grade potassium iodide in the quantities needed.

Saturday morning, shortly after 3:00 a.m., the Mallinckrodt Chemical Company agreed to provide HEW with approximately a quarter million one-ounce bottles of the drug. Mallinckrodt in St. Louis, working with Parke-Davis in Detroit and a bottle-dropper manufacturer in New Jersey, began an around-the-clock effort. The first shipment of potassium iodide reached Harrisburg about 1:30 a.m. Sunday. By the time the last shipment arrived on Wednesday, April 4, the supply totaled 237,013 bottles.

SATURDAY MARCH 31

The great concern about a potential hydrogen explosion inside the TMI-2 reactor came with the weekend. That it was a groundless fear, an unfortunate error, never penetrated the public consciousness afterward, partly because the NRC made no effort to inform the public it had erred.

Around 9:30 p.m. Friday night, the NRC chairman asked Roger Mattson to explore the rate at which oxygen was being generated inside the TMI-2 reactor system and the risk of a hydrogen explosion. "He said he had done calculations," Mattson said in his deposition. "He was concerned with the answers." Mattson is director of the Division of Systems Safety within the Office of Nuclear Reactor Regulation (NRR), which is headed by Denton, and had spent part of Thursday and Friday working on how to remove a gas bubble from the reactor. Following Denton's departure for TMI, Mattson served variously as NRR's representative or deputy representative at the Incident Response Center.

Hydrogen had been produced in the reactor as a result of a high-temperature reaction that occurred between hot steam and the zirconium cladding of the fuel rods. For this hydrogen to explode or burn -- a less dangerous possibility -- enough oxygen would have to enter the system to form an explosive mixture. There were fears this would happen as the result of radiolysis. In this process, radiation. breaks apart water molecules, which contain hydrogen and oxygen.

Two NRC teams worked throughout the weekend on the problem, and both sought help from laboratories and scientists outside the NRC. One group addressed the rate at which radiolysis would generate oxygen at TMI-2. The second analyzed the potential for hydrogen combustion. Robert Budnitz of the

NRC also asked experts about possible chemicals that might remove the hydrogen.

At noon, Hendrie talked by telephone with Denton and expressed his concern that oxygen freed by radiolysis was building up in the reactor. Earlier, Hendrie had told Victor Stello, Jr., Denton's second-in-command at TMI, the same thing. The NRC chairman told Denton that Governor Thornburgh should be made aware of the potential danger. Denton promised to speak with Thornburgh.

Shortly after 1:00 p.m., Mattson got some preliminary answers regarding the potential for a hydrogen explosion. An hour later, Mattson got more replies. "I had an estimate there was oxygen being generated, from four independent sources, all with known credentials in this field," he said in his deposition. "The estimate of how much oxygen varied, but all estimates said there was considerable time, a matter of several days, before there was a potential combustible mixture in the reactor coolant system."

At a Commission hearing, Mattson later admitted in response to questions from Commissioner Pigford that the NRC could have determined from the information available at that time that no excess oxygen was being generated and there was no real danger of explosion.

But when Mattson met with the NRC commissioners at 3:27 p.m. on Saturday, "the bottom line of that conversation . . . was there were several days required to reach the flammability limit, although there was oxygen being generated," Mattson recalled in his deposition. "And I expressed confidence that we were not underestimating the reactor coolant system explosion potential; that is, the estimate of 2 to 3 days before reaching the flammability limit was a conservative estimate." By Saturday night, however, Mattson would be told by his consultants that their calculations indicated that the oxygen percentage of the bubble was on the threshold of the flammability limit.

Around 6:45 p.m., Mattson talked with Vincent Noonan, the man within NRC most knowledgeable about what might result from an explosion inside a reactor. One NRC consultant had predicted that a hydrogen blast would produce pressures of 20,000 pounds per square inch within the TMI-2 reactor. B&W, designer of the reactor, however, had considered the dampening effects of water vapor on an explosion and those of an enriched hydrogen environment and calculated a total pressure of 3,000 to 4,000 psi. That was encouraging. Previous analyses indicated the reactor coolant system of a TMI-2 reactor could withstand blast pressures of that magnitude.

Late Saturday evening, James Taylor of B&W reiterated another B&W engineer's conclusion first relayed to the NRC Thursday night -that no excess oxygen was being generated. That information, Mattson stated in his deposition, never reached him.

Saturday at 2:45 p.m., Hendrie met with reporters in Bethesda. He said then that a precautionary evacuation out to 10 or 20 miles from the Island might be necessary if engineers attempted to force the bubble out of the reactor. NRC had concluded such an attempt might cause further damage to the core, Hendrie said, and it might touch off an explosion of the bubble.

Stan Benjamin, a reporter with the Washington bureau of the Associated Press, followed up Hendrie's press conference by interviewing two NRC officials: Edson Case, Denton's deputy in the Office of Nuclear Reactor Regulation, and Frank Ingram, a public information spokesman. From them, and an NRC source he refused to name, Benjamin learned of the concern within the Incident Response Center that the bubble could become a potentially explosive mixture within a matter of days, perhaps as few as two. Benjamin checked his story with Case and Ingram, reading much of it to them word by word, before releasing the article. Case and Ingram agreed it was accurate. The report -- first transmitted as an editor's note at 8:23 p.m. -- was the first notice to the public that some NRC official feared the bubble might possibly explode spontaneously.

Denton had been briefed throughout Saturday afternoon and evening by Hendrie and NRC officials in Bethesda on the oxygen estimates and the potential for a burn or explosion. But he learned of the AP story only a short time before he joined Governor Thornburgh and Lieutenant Governor Scranton for a late evening press conference in Harrisburg. The Governor assured reporters that "there is no imminent catastrophic event foreseeable at the Three Mile Island facility." Denton, too, said: "There is not a combustible mixture in the containment or in the reactor vessel. And there is no near-to danger at all." Denton also tried to deflate the impression, voiced by several reporters, that contradictions existed between himself and his colleagues at NRC headquarters. "No, there is no disagreement. I guess it is the way things get presented," he said.

But there was disagreement, and Denton wanted it resolved. President Carter had announced earlier in the evening he would visit TMI the following day. Denton told Stello to explore the oxygen-hydrogen issue further with outside experts. Stello realized the concern in Washington. He had received a telephone call shortly after 9:00 p.m. from Eugene Eidenberg, a Presidential aide, inquiring about the AP story. Stello told the White House that he did not share the concern felt at NRC headquarters.

***

Saturday, as the NRC wrestled with managing the accident and the envisioned danger of the hydrogen bubble, officials of the Department of Health, Education, and Welfare struggled with their own concerns. That morning, senior HEW health officials gathered and continued the previous day's discussion of the possibility of an evacuation; for the first time, they debated how large an area should be evacuated. But the discussions led ultimately to a recommendation to consider immediate evacuation if the NRC could not provide assurances that the reactor was cooling safely. Joseph Califano HEW Secretary, summarized the group's views in a memorandum to Jack Watson of the President's staff.

Later in the day, HEW health officials attended an interagency meeting at the White House, convened by Watson, and repeated the HEW recommendation to consider evacuation. Richard Cotton, a key Califano aide, raised another Califano recommendation that NRC officials consult with HEW and Environmental Protection Agency experts regarding; the potential health effects of the efforts to control TMI-2's reactor. Cotton persisted after the meeting, and on Sunday and Tuesday HEW officials were briefed by the NRC. These briefings, however, were always informational; there was no NRC effort to seek HEW's advice.

SUNDAY, APRIL 1

Throughout Saturday night and the early hours of Sunday, county emergency preparedness offices were deluged with telephone calls from citizens concerned by the conflicting reports about the hydrogen bubble. But the flow of useful information from the state to the local level had essentially ceased after Denton's arrival. The Governor's office focused attention on the federal effort -- Denton and officials from several U.S. emergency agencies. Oran Henderson, director of the Pennsylvania Emergency Management Agency, was no longer invited to the Governor's briefings and press conferences, and he did not attend after Friday night. Thus PEMA – although it continued to receive status reports from the Bureau of Radiation Protection -- was isolated from information wanted at the local level.

In Dauphin County, frustration ran high. Shortly before midnight on Saturday, State Sen. George Gekas called the Governor in an attempt to obtain accurate information. Gekas was told the Governor was too busy to talk. Then Gekas called Scranton, and got the same response. At that point, Gekas told a Scranton aide that unless more cooperation and information were forthcoming, Dauphin County would order an evacuation at 9:00 a.m. Sunday. Scranton called the county's emergency center at 2:00 a.m. and agreed to meet officials there later in the morning. The Lieutenant Governor arrived at 10:00 a.m., preceded by Henderson, who complained of his own inability to obtain information. Scranton listened to Molloy and his colleagues.

"I think he was just totally shocked by what was transpiring at our level; how busy we were; how much work we were doing; how complicated it was," Molloy said in his deposition.

***

Sunday, Mattson and several other NRC staffers met with NRC Commissioners Hendrie, Victor Gilinsky, and Richard Kennedy. Their purpose was to reach a judgment, based on the estimates and information available, about the true potential for a hydrogen explosion inside the reactor.

According to Mattson's deposition, the group agreed: 5 percent oxygen was a realistic flammability limit, 11 percent oxygen was a realistic detonation limit, that there could be no spontaneous combustion below 900°F, that the oxygen production rate was approximately one percent per day, and that the present oxygen concentration in the bubble was 5 percent.

After the meeting, Hendrie and Mattson drove to TMI to meet with Denton.

Stello talked with Denton Sunday morning and outlined his arguments against any danger of a hydrogen explosion inside the reactor. Pressurized water reactors, the type used at TMI-2, normally operate with some free hydrogen in the reactor coolant. This hydrogen joins with the oxygen freed by radiolysis to form another water molecule, which prevents the build-up of oxygen to a quantity that would allow an explosion to take place. Stello told Denton that the process was the same now, and there was no danger of explosion.

Hendrie and Mattson met with Denton and Stello in a hangar at Harrisburg International Airport minutes before President Carter's 1:00 p.m. arrival. Mattson and Stello had not talked to each other since Friday morning. Mattson outlined the conclusions reached at NRC headquarters about the bubble and the reasoning behind them. In an interview with the Commission staff, Mattson described what happened next:

"And Stello tells me I am crazy, that he doesn't believe it, that he thinks we've made an error in the rate of calculation . . . . Stello says we're nuts and poor Harold is there, he's got to meet with the President in 5 minutes and tell it like it is. And here he is. His two experts are not together. One comes armed to the teeth with all these national laboratories and Navy reactor people and high faluting PhDs around the country, saying this is what it is and this is his best summary. And his other [the operating reactors division] director, saying, 'I don't believe it. I can't prove it yet, but I don't believe it. I think it's wrong.' "

Upon the President's arrival, Denton briefed the Chief Executive on the status of the plant and the uncertainty regarding its infamous bubble.

The President was driven to TMI, put on protective yellow plastic shoecovers, and toured the facility with Mrs. Carter, Governor Thornburgh, and Denton. Stello, Hendrie, and Mattson went to the temporary NRC offices. During the afternoon, experts -- including those at Westinghouse and General Electric -- were canvassed by phone. "By three o'clock, we're convinced we've got it," Mattson said in his interview. "It's not going to go boom."

NRC scientists in Bethesda eventually reached the same conclusion, but later in the day. Shortly before 4:00 p.m., NRC Commissioners Richard Kennedy, Peter Bradford, and John Ahearne met. They expressed concern over the differing estimates presented by the NRC staff and decided there might be a need to consider evacuation.

Kennedy telephoned Hendrie at TMI and told him the three NRC Commissioners thought Governor Thornburgh should advise a precautional evacuation within 2 miles of the plant, unless experts on-site had better technical information than that available in Bethesda. Hendrie assured Kennedy that the free hydrogen inside the reactor would capture any oxygen generated and that no problem existed.

In midafternoon, new measurements showed the large bubble in the reactor was diminishing. The gases still existed, but they were distributed throughout the system in smaller bubbles that made eliminating the predominantly hydrogen mixture easier. Why this occurred, no one knows. But it was not because of any intentional manipulation by Met Ed or NRC engineers.

By late Sunday afternoon, NRC -- which was responsible for the concern that the bubble might explode -- knew there was no danger of a blast and that the bubble appeared to be diminishing. It was good news, but good news unshared with the public. Throughout Sunday, the NRC made no announcement that it had erred in its calculations or that no threat of an explosion existed. Governor Thornburgh was not told of the NRC miscalculation either. 'Nor did the NRC reveal the bubble was disappearing that day, partly because the NRC experts themselves were not absolutely certain.

MONDAY, APRIL 2

Monday morning Denton and Mattson met the press. George Troffer, a Met Ed official, had already told a reporter the bubble was essentially gone. Denton acknowledged a "dramatic decrease in bubble size," but cautioned that more sophisticated analyses were needed "to be sure that the equations that are used to calculate bubble size properly include all effects." As to the bubble's potential for explosion, Denton told reporters "the oxygen generation rate that I was assuming yesterday when I was reporting on the potential detonation inside the vessel is, it now appears, to have been too conservative." Throughout the press conference, Denton continued to refer to NRC's estimates as too conservative; he never stated outright that NRC had erred in its conclusion that the bubble was near the dangerous point.

According to Mattson, the tone of the press conference -- its vagueness and imprecision -- was decided upon at a meeting of NRC officials Monday morning.

> We wanted to go slow on saying it was good news. We wanted to say it is good news, do not panic, we think we have got it under control, things look better, but we did not want to firmly and finally conclude that there was no problem. We had to save some wiggle room in order to preserve credibility. That was our judgement."

Bibliography

I. OUTSTANDING

Paul Craig and John Jungerman, **Nuclear Arms Race** McGraw-Hill, New York 1986. In spite of its title, this book is the best source of readable and understandable information about physics fundamentals and technology of nuclear systems and the medical and biological effects of radioactivity. By two researchers who had worked at Los Alamos, where the first nuclear bombs were built. Written while the cold war was still going strong, there is a strong emphasis on weapons, and a wealth of data on nuclear arsenals and destructive power of various warheads, with an intelligence and sensitivity to the needs of both the humanities-oriented and the scientifically-oriented reader, it serves both very well. There is unfortunately very little about reactors, but there are two other *Outstanding* books that can make up for that.

Anthony V. Nero, **A Guidebook to Nuclear Reactors**, University of California Press, Berkeley, California, 1979. Here is a book that is what it says it is. It is *all* about reactors. No math, but lots of numbers, graphs, and diagrams. Not a book for cover-to-cover reading, but a well organized reference. Written in an academic style, well documented, a little hard to get through because of its thoroughness.

Richard Wolfson, **Nuclear Choices, A Citizen's Guide to Nuclear Technology**, MIT Press, Cambridge, Massachusetts, 1991. Peppered with interesting "Nuclear News" Boxes, this book is comprehensive, well written, and at a popular level. It is based on an elementary college course taught by the author, and deals about as well as it is possible in a purely verbal treatment. The Choices part of the book come through in political statements quoted from various sources, but what the choices are is not the main message of the book.

II. TEXTS

Fujiya Yang and Joseph Hamilton, **Modern Atomic and Nuclear Physics**, McGraw-Hill, NY, 1996. Perhaps the best recent book for the serious student with moderate mathematical skills. Combines abstraction with excellent explanation for the beginner. With facts and figures, well chosen for relevance.

T.A. Littlefield and N. Thorley, **Atomic and Nuclear Physics**, Van Nostrand Reinhold, New York, 1979. A good basic text, not for the general reader, but accessible for someone with questions.

David Halliday, **Introductory Nuclear Physics**, John Wiley & Sons, New York, 1955. The basic physics of the nucleus, by the author of the most used Introductory Engineering Physics Text. Mathematical and conceptual at a high level, it is for someone planning to go into this field. All the basic material, surprisingly still thoroughly appropriate, typically well written, and useable.

W.N. Cottingham, **An Introduction to Nuclear Physics**, Cambridge University Press, 2001. Not available for review.

III. GENERAL

George Kemeny, chairman of the Commission,**The Report of the President's Commission on the Three Mile Island Accident** U.S. Government Printing Office, Washington, DC, 1979. A dry and carefully crafted compromise document, which nevertheless carries the punch that George Kemeny and a few others on the Commission were able to give it. The Commission was designed to be representative of industry, government, academia, and public. Interesting, but not easy reading. The high point is the surprisingly well written diary of the minute by minute doings in and around the reactor during ten days of crisis. The account is in the public domain, and is reprinted in the Appendix of this book for its timelessness and its penetrating insights on the human as well as technical side of that particular event. The book is available at many research libraries. Not all the copies include the "Account of the Accident." We obtained our copy at the Pennsylvania State Library in Harrisburg, PA.

IV COMPREHENSIVE

James Duderstadt and Chihiro Kikuchi, **Nuclear Power – Technology on Trial**, University of Michigan Press, Ann Arbor, Michigan, 1979. More historical and more detailed than most books, it starts off with a section titled, "Nuclear Power: Necessary or Not?" Typically, it gives a balanced answer. We can't stop using it, but, on the other hand, and then on still another hand are "perceived hazards ... but they are hypothetical." The book is pedantic, but authoritative; packed full of information and insightful if indecisive commentary.

V. OPINION

Bernard Cohen, **Before It's Too Late**, Plenum Press, New York, 1983. Written by a scientist who describes it as an attempt –"for once – to get the viewpoint of the main-line scientific community across to the public," it is unabashedly opinionated, and a response to the "phobic fear of radiation," that has been spread by "fringe types." It is not very scientific, but is worthwhile reading because it makes the pro-nuclear case that has held sway in policy circles and in the industry, if not in the public.

Michio Kaku and Jennifer Trainer, **Nuclear Power: Both Sides**, W. W. Norton & Co., New York, 1982. This is a collection of pieces that come under the advertised aim, "Are you pro- or anti-nuclear? Here is a book that will help you decide ... Experts from both sides of the controversy." If you look at this as a yes or no issue, you will find it interesting to see the most vociferous debaters on the issues take each other on. More like a political debate, the quality of the articles varies from irresponsibly biased to well informed and reasonably argued. Unfortunately opinion articles leave a credibility gap that limits the enlightenment, and leaves the impression that the citizen reader makes up his or her mind based on choosing the most persuasive experts.

John G. Fuller, **We Almost Lost Detroit,** Readers Digest Press, 1975. The gripping story of the near-disaster at Fermi I, the first reactor at the Monroe, Michigan site.

John J. Berger, **Nuclear Power, the Unviable Option**, Ramparts Press, Palo Alto, California, 1976. A strong case that it's too costly, too dangerous, and not necessary.

Index

Account of the Accident 215
  radiation levels 228
  Standard Review Plan 231
  Babcock & Wilcox 219
  bubble, hydrogen 259
  Captain Dave 233
  control rods 216
  control room 218
  core damage 245
  core uncovery 237
  dumping wastewater 245
  ECCS 232
  evacuation 251
  feedwater 216
  Harrisburg Mayor 234
  HPI 220
  LOCA 217
  Loss Of Coolant 217
  maintenance tag 217
  Met Ed 219
  Nuclear Regulatory Commission . 229
  polisher 219
  PORV 216
  pressurizer 216
  radiation 244
  radioactive iodine 240
  scram 216
  site emergency 228
  Standard Review 231
  Susquehanna River 244
  thyroid 240
  WKBO 233
  women and children 251
acid and base 18
activation 66
Aldama Street 202
American Nuclear Society 204
Avogadro's number 40
Baker, Russel v
Bartlett, A.A. 143
Becquerel, Henri 159
Berger, John J. 272
Bibliography 269
binding energy 37, 42
  graph 45
boiling water reactors 120
bomb 81
bond . . . . i, 18
  between ions 18
breeder reactors 136
cancer 187
  treatment 195
carbon-14 167
  dating 169
centrifugation 82
Chadwick, James 69
chain reaction 75
chemical i
chemical energy 5
Chernobyl 131
China syndrome 207
Cohen, Bernard 271
condenser 122
confinement 141
  inertial 141
  magnetic 142
containment building 110
control rods 114
conversion 3
convert i
cookie sheets 26
coolant
  primary 113
cooling towers 123
core 108
Coulomb 19
Craig, Paul 269
Critical mass 76
Curie, Marie and Pierre 159
deuterium 38

diffusion 82
DNA 186
downwinders 147
Duderstadt, James 271
$E=mc^2$ 3
Einstein . . . . ii, 12
electric field 19
  energy density 28
electron 23
emergency FW pumps 126
energy . . . . i, 16
energy barrier 51, 66
energy yield
  uranium fission 72
enrichment 81
equivalence of mass and energy i
Excess Reactivity 102
  Temperature Coefficient 104
experts iv
exponential
  decrease 157
  process 157
exposure 194
  REM 182
feedback mechanism 99
Feedwater Pumps 126
fission 54
Force-Work-Energy iii
fossils 170
Frisch, Otto 70
fuel . . . . i, 110
  fuel rods 110
Fuller, John G. 272
fusion 60, 139
gasoline 6
Geiger tube 181
generation 85
generation ratio 76, 87
  greater than one 94
  less than one 92
generation time 86
Goldring, Hanna i
graphite 130
gravitational bond 6
Habicht, Conrad 12
Hahn, Otto 69
half life 151, 152, 161
Halliday, David 270
Hamilton, Joseph 270
hats 153
heat exchanger 121
heavy water 38, 129
high pressure injection 126
Hiroshima i
isotopes 35
Joule 73
Jungerman, John 269
Kaku, Michio 272
Kemeny, George 211, 270
Kikuchi, Chihiro 271
Kilowatt Hours 73
Lapland
  reindeer 193
Littlefield, T.A. 270
mass . . . . i, 16
mass and energy 15
mass defect 10, 11
  energy defect 12
matter i
Meitner, Lise 70
meltdown 209
moderator 101, 112
Nero, Anthony 269
neutralization 19
nuclear bonds 34
nuclear force i
nuclear fuel 51
Nuclear Society 204
nucleon 9
people-REM 191
periodic table 10
photon 24
Pilot Operated Relief Valve 126

plutonium
reactor 133
Polisher 127
PORV 126
President's Commission 270
pressurizer 109
probability density 24
properties 16
proton 23
public interest iv
radiation 180
Birth Defects 189
Delayed effects 186
Immediate Effects 184
radiation sickness 195
radioactive isotopes
table 164
radioactive particles 149
alpha 149, 173
beta 149, 173
gamma 149, 173
radioactivity 145, 160
exposure 179
radium 170
radon 172, 174
reaction
fission 71
reactor 108
pressurized-water 108
schematic 119
rearrangement 58
rearrangements 49
refueling 118
risk
assessment 199, 205
saturation 50
SCRAM 115, 208
secondary coolant 121
shim adjustment 117
size 9
source of nuclear energy 1
spent fuel 110
stability 78
statistics 206
steam generator 121
Strassman, Fritz 69
strong nuclear force 9
Summary of Part I 65
sump pump 126
Takahama 201
teachers
science 204
thermal neutrons 101
Thorley, N. 270
Three Mile Island 210
Tokamak 142
Trainer, Jennifer 272
tritium 46
uranium 53
dissociation sequence 171
watts 73
weight 16
Wolfson, Richard 269
Work–Energy Theorem 17
Yang, Fujiya 270